Automated Vehicles Symposium 2015

Ann Arbor, Michigan, USA
21 - 23 July 2015

ISBN: 978-1-5108-5182-5

Printed from e-media with permission by:

Curran Associates, Inc.
57 Morehouse Lane
Red Hook, NY 12571

proceedings.com

Some format issues inherent in the e-media version may also appear in this print version.

Printed by Curran Associates, Inc. (2018)

For permission requests, please contact Association for Unmanned Vehicle Systems International at the address below.

Association for Unmanned Vehicle Systems International
2700 South Quincy Street
Suite 400
Arlington, VA 22206
USA

Phone: (703) 845-9671
Fax: (703) 845-9679

www.auvsi.org

Additional copies of this publication are available from:

Curran Associates, Inc.
57 Morehouse Lane
Red Hook, NY 12571 USA
Phone: 845-758-0400
Fax: 845-758-2633
Email: curran@proceedings.com
Web: www.proceedings.com

TABLE OF CONTENTS

Reducing accidents with today's active safety technologies

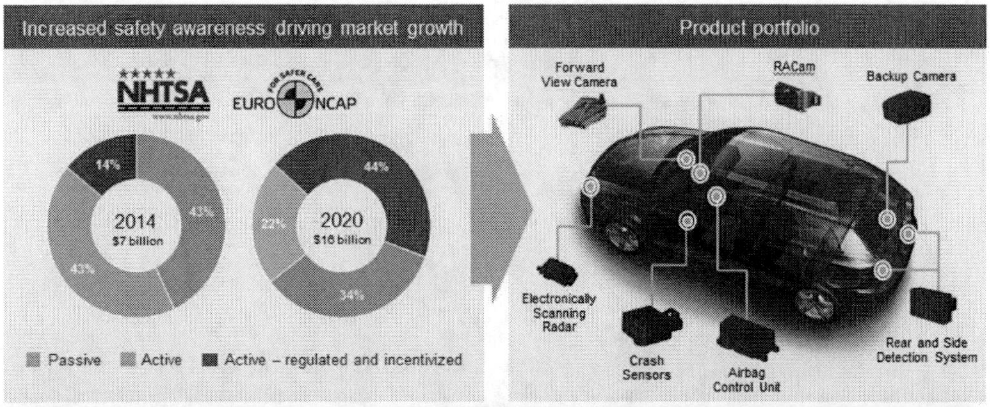

Integrated technology advances Active Safety value proposition

DELPHI

NHTSA automated driving levels

	Level 0	Level 1	Level 2	Level 3	Level 4
	Driver only	Assisted	Partial	Conditional	Full
Feature		Active high beam Collision imminent braking Cruise control	Traffic jam assist Adaptive cruise & lane keeping Self-parking (with driver)	Collision avoidance Automated highway Automated urban	Valet self-parking Highway point-to-point Urban point-to-point
Technology		Radar Forward sensors	Radar Forward sensors Multi-domain controller Driver state sensor V2X	LIDAR & 360° radar High accuracy GPS Multi-domain controller Forward, HD & IR cameras V2X Internal moment unit	LIDAR & 360° radar High accuracy GPS Multi-domain controller Forward, HD & IR cameras V2X Internal moment unit
		Today	Today	2020	2025+

DELPHI

3

The road to automated driving

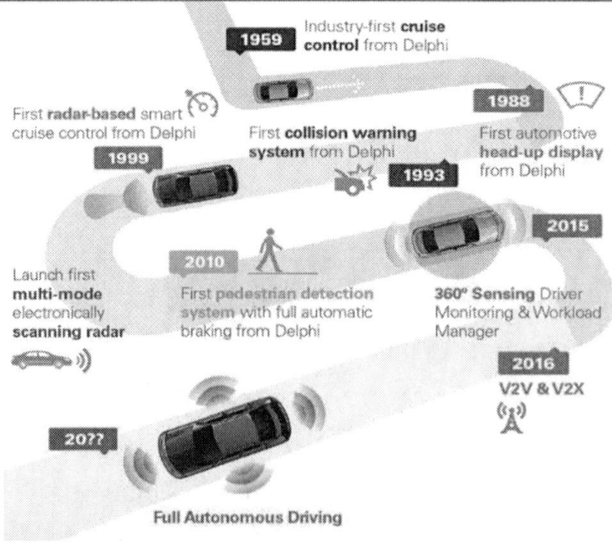

4

Delphi completed first coast-to-coast trip in an automated vehicle

DELPHI

Re-inventing the roadtrip

Automated drive highlights	Arriving in New York City
~3,400 miles traveled	
~99% of trip fully automated	
9 days, 15 states	
20 sensors	
~3 terabytes of data collected	

Setting records and breaking boundaries

DELPHI

Automated driving video

Video Presentation in
Original e-media

DELPHI

What we learned

Key findings

- Able to navigate multiple lanes of traffic - up to 6 in high congestion areas
- Multiple sensors are required
  - Radars performed well in all conditions
  - Vision performed well in most conditions
- Possible to operate in automated mode nearly 100% of the time on US Highways

Next steps…

- Further develop cameras to detect diverse lane markings and signage
- Improve camera performance in bright sunlight and harsh weather conditions
- Driving behavior optimization for a more natural feel
- Democratization of Automated Driving features for production programs

Further improving the technology of the future

DELPHI

Scenarios Encountered

DELPHI

DELPHI

Innovation for the Real World

Cars - the better drivers ?

Dr. Kay Stepper

Vice President, Driver Assistance and Automated Driving

1 BOSCH

Vision – Injury free driving

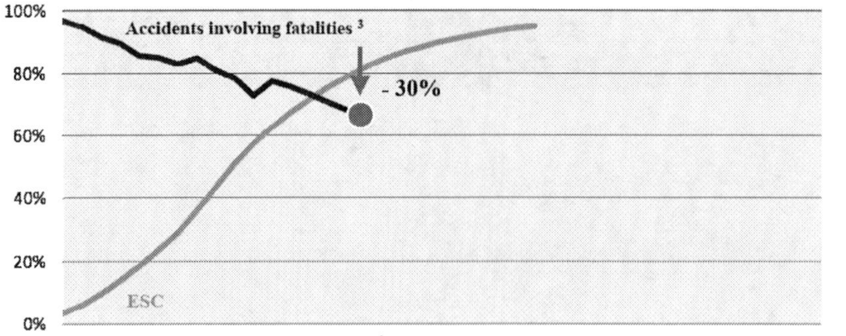

2 Source: Bosch Accident Research, DAT, DESTATIS [1] Installation rate and figures estimated [3] 100% in 1999 = 395 689 accidents w/ casualties

BOSCH

Nur Zusatzhinweise

BOSCH

Vision – Injury free driving

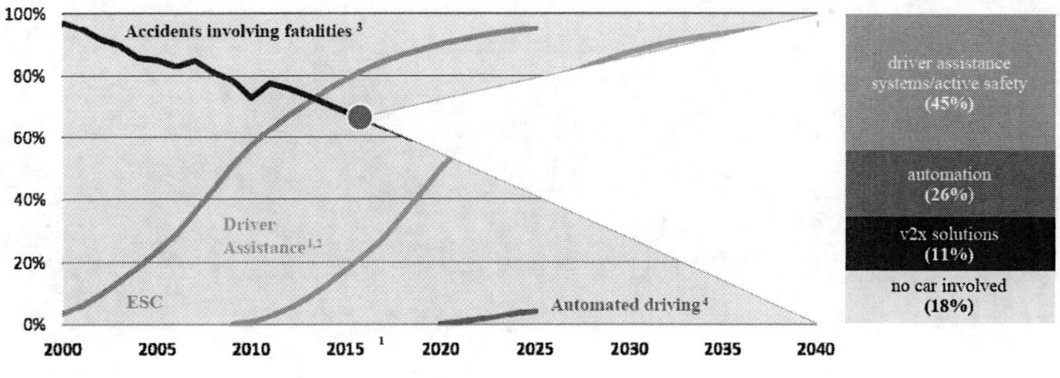

Source: Bosch Accident Research, DAT, DESTATIS [1] Installation rate and figures estimated [2] AEB including AEB VRU, LDW/LKS, Night Vision, Lane Change Assist (incl. BSD), Crossing Assist, Left Turn Assist [3] 100% in 1999 = 395 689 accidents w/ casualties [4] possible Installation rate of automated function e.g. Highway Pilot

BOSCH

The human driver

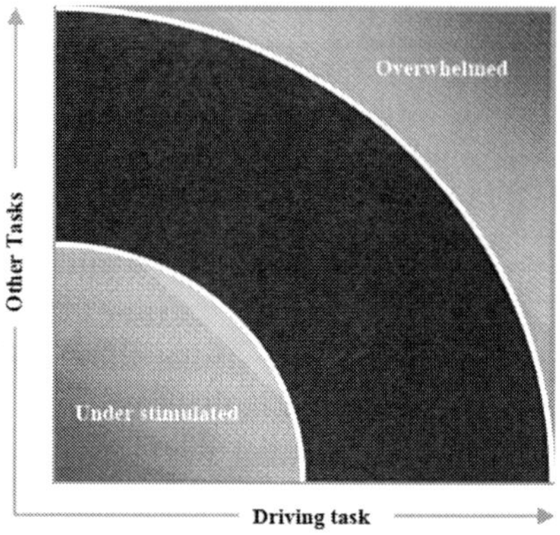

5 BOSCH

Gradual progress toward automated driving

6 BOSCH

From driver to passenger

7

Assisted driving

8

Assisted driving

9 BOSCH

Example - Evasive steering support

10 BOSCH

Notes

11

BOSCH

Partial / Conditional Automation

12

BOSCH

High / Full Automation

13 BOSCH

Key technologies for automated driving

Perception 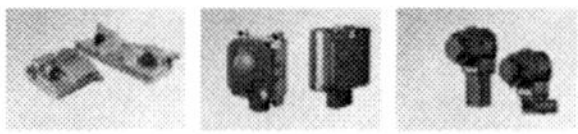

Localization 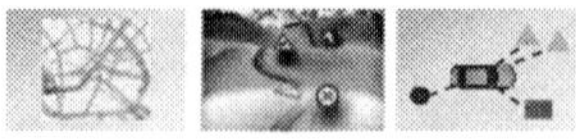

Planning & Decision Making 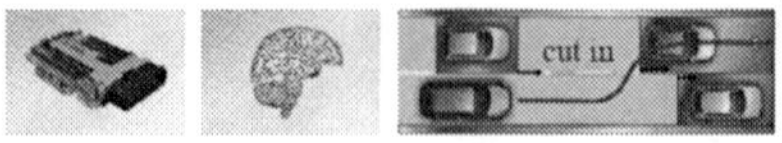

Vehicle motion control 

Driver monitoring 

14 BOSCH

The car takes over fallback responsibilities

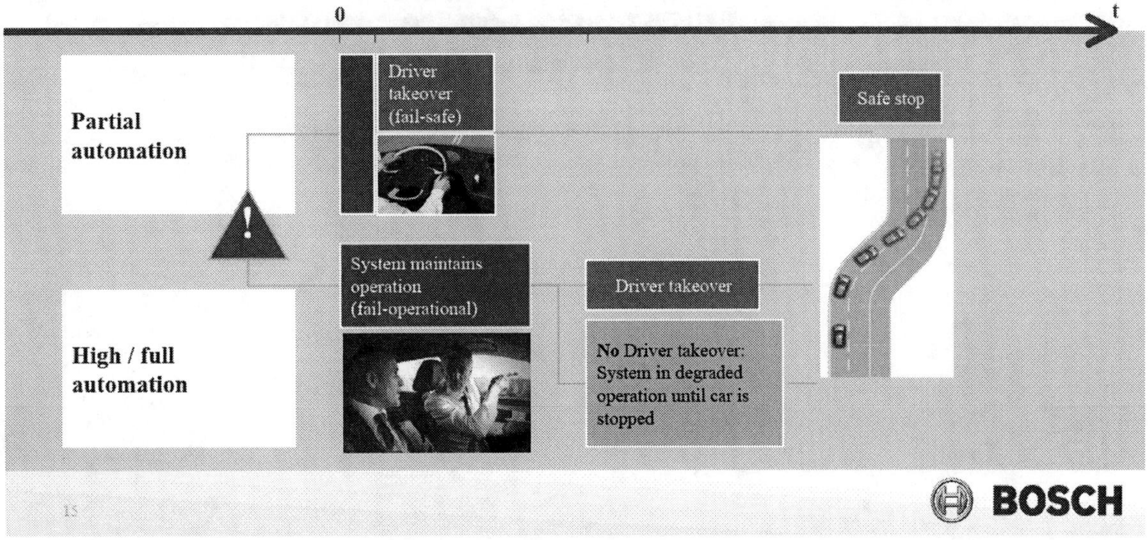

Highly automated driving requires redundancies

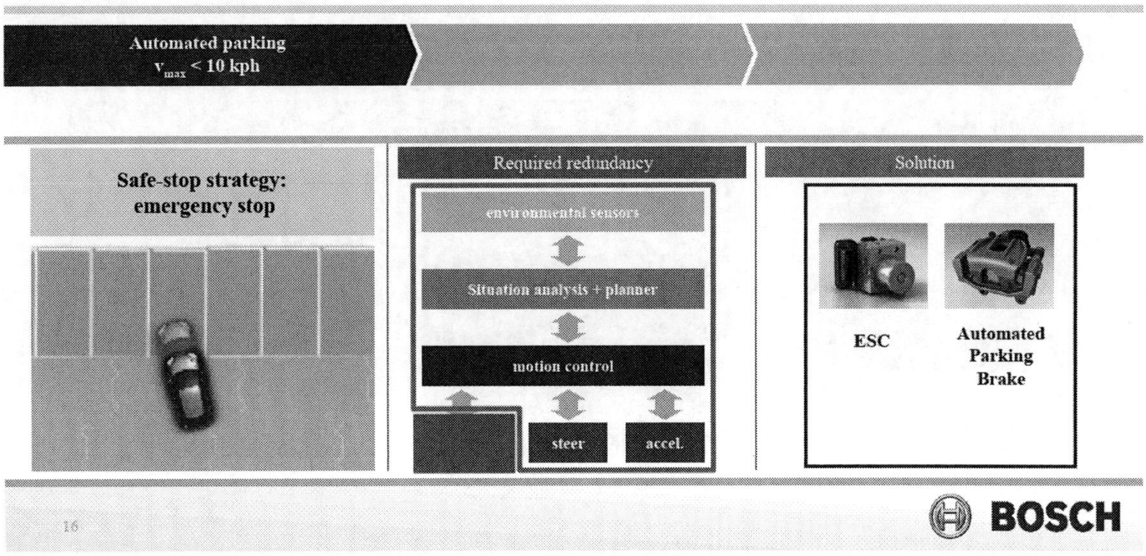

Highly automated driving requires redundancies

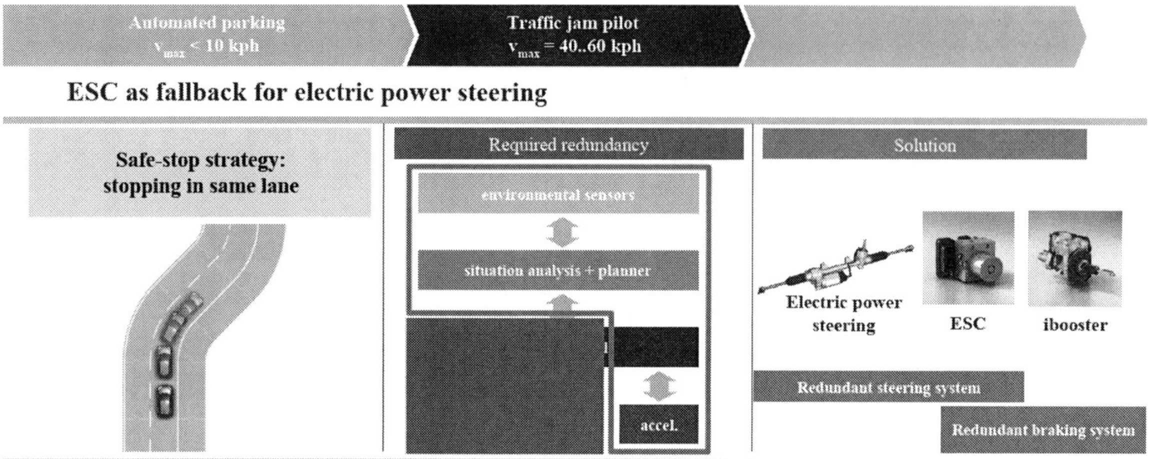

17

Highly automated driving requires redundancies

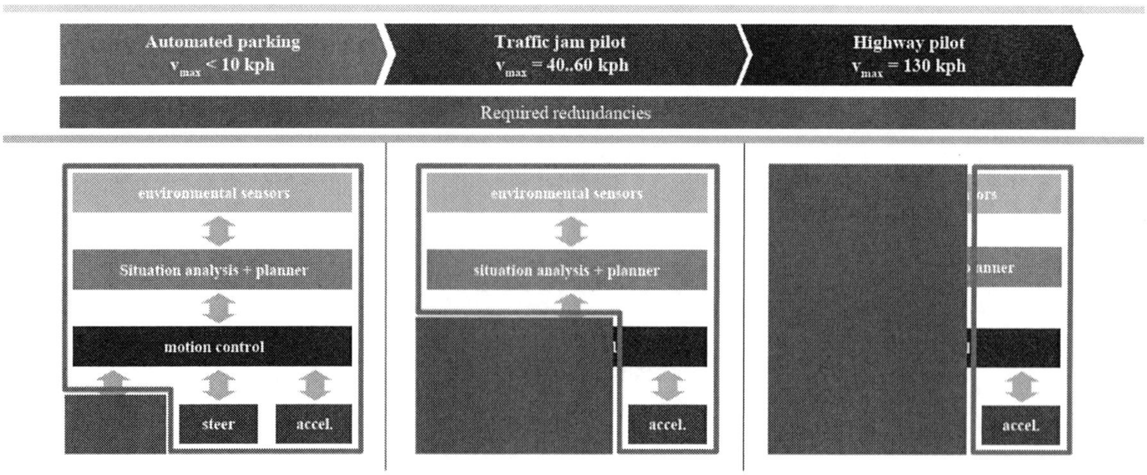

18

Test, Verify, Validate … Repeat

Connectivity based development and validation

Example

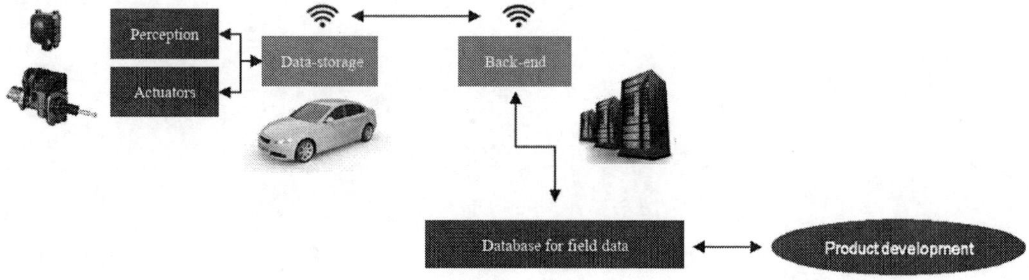

Connected vehicles are better vehicles

Small latency
(e.g. electronic brake light)

Non-visible area
(e.g. crossing traffic)

Information on future behavior
(e.g. remaining red light phase)

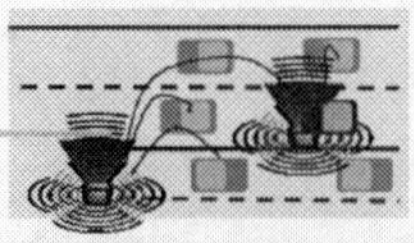

Redundancy
(360° vehicle environment)

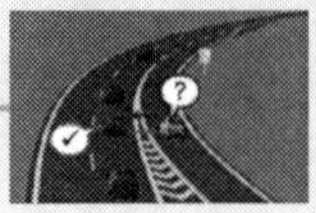

Negotiation of maneuvers
(lane merge assist)

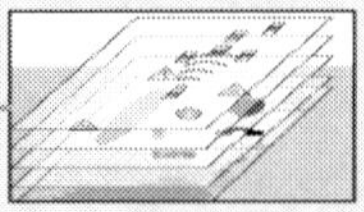

Sensor data fusion
(e.g. dynamic map)

BOSCH

Notes

BOSCH

Cars - the better drivers ?

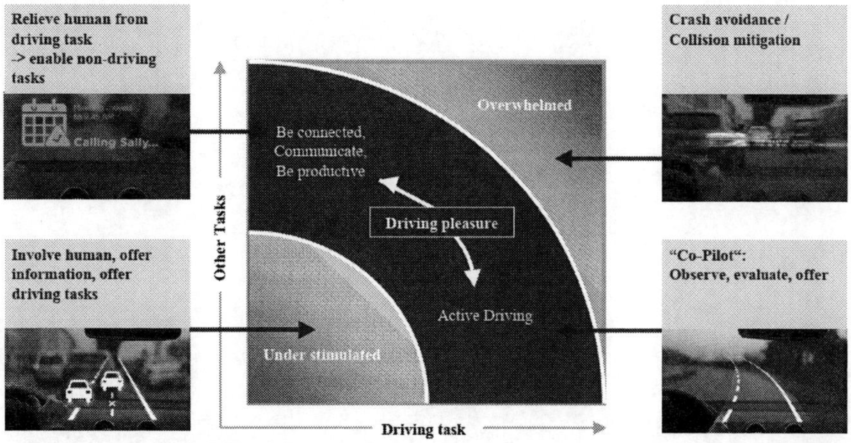

23 BOSCH

Vision – injury free driving

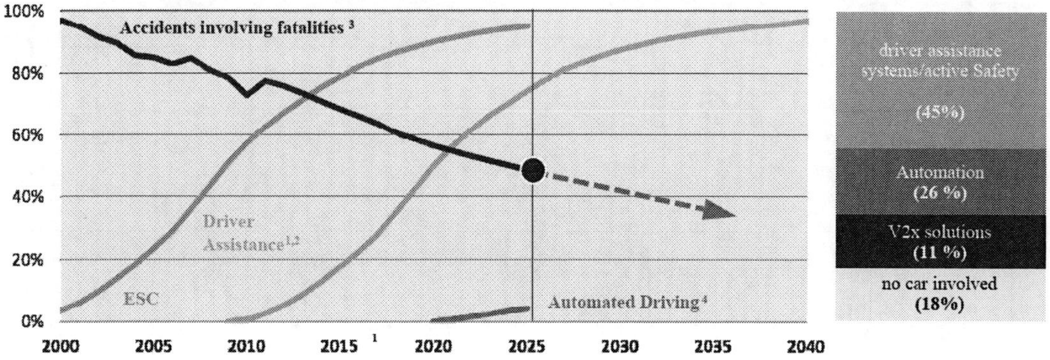

24 Source: Bosch Accident Research, DAT, DESTATIS [1] Installation rate and figures estimated [2] AEB including AEB VRU, LDW/LKS, Night Vision, Lane Change Assist (incl. BSD), Crossing Assist, Left Turn Assist [3] 100% in 1999 = 395 689 accidents w/ casualties

BOSCH

Invented for life

25 BOSCH

DAIMLER

Andreas Knapp
Daimler AG, Germany

Legal issues addressed in the EU funded AdaptIVe project

AVS 2015
Ann Arbor, USA
21 July 2015

Agenda

- Project facts
- Motivation
- Challenges, objectives, and concept

Facts

Duration:	42 months (January 2014 - June 2017)
Budget:	EUR 25 Million
European Commission:	EUR 14,3 Million
Coordinator:	Aria Etemad, Volkswagen Group Research
8 Countries:	France, Germany, Greece, Italy, Spain, Sweden, The Netherlands, United Kingdom

29 partners

The integrated project AdaptIVe

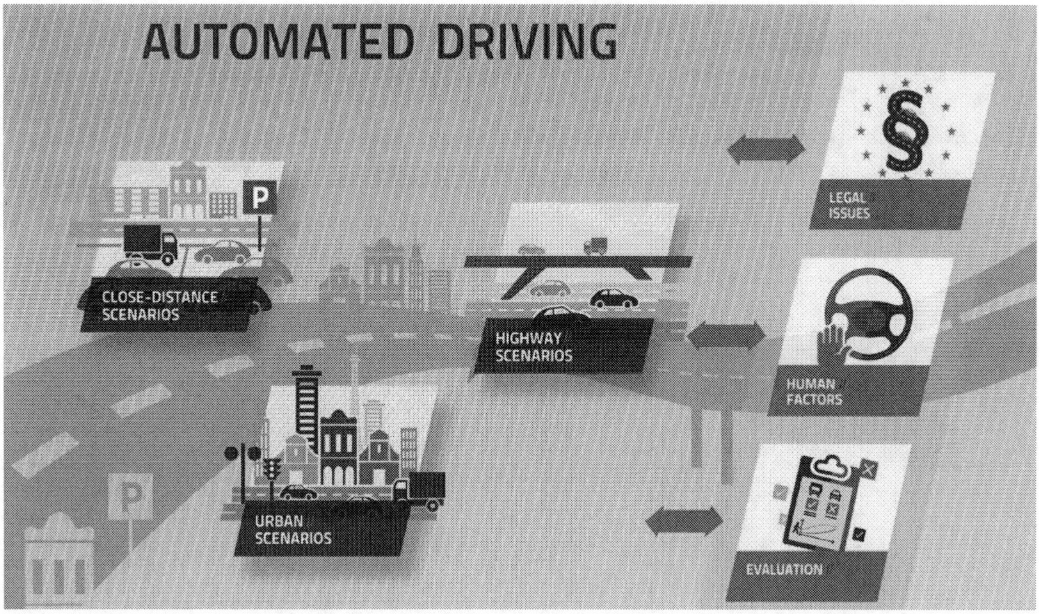

Agenda

- Project facts
- Motivation
- Challenges, objectives, and concept

Potentials for automated driving

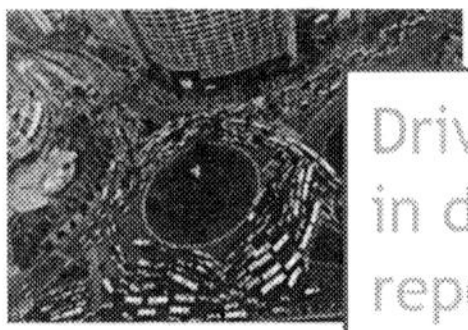

Drivers are supported in demanding or repetitive tasks. Travel comfort increases.

Vehicles dynamically adapt the level of automation according to the current situation.

Vehicles react more effectively to external threats.

Vehicles are resilient to different types of system and human failure.

Motivation for automated driving functions

Zero emission	Reduction of fuel consumption & CO_2 emission Optimization of traffic flow	
Demographi c change	Support unconfident drivers Enhance mobility for elderly people	
Vision zero	Potential for more driver support by avoiding human driving errors	

Demonstrators

Actuators, key fob, ultrasonic and radar sensors, cameras, on-spot mapping

Actuators for vehicle control, laser scanner and radar sensors, cameras

Actuators, ECUs, on-board sensors, radar, map-based electronic horizon, V2X

Challenges and project objectives

Widespread application of automated driving to improve traffic safety, efficiency and comfort

Response 4

Legal issues - Response 4

Levels of driving automation

level 0	level 1	level 2	level 3	level 4	level 5
	LKA ACC	Parking Assistance	Traffic Jam Chauffeur	Parking Garage Pilot	
No auto-mation	Assisted	Partial auto-mation	Condi-tional auto-mation	High auto-mation	Full auto-mation

Driver in the loop

- No significant change with respect to existing driver assistance systems

Driver out of the loop

- Not in accordance with regulatory law (Vienna Convention of 1968, national road law)
- Shared responsibility for control between driver and system
  → need for action

Source: SAE document J3016, "Taxonomy and Definitions for Terms Related to On-Road Automated Motor Vehicles", issued 2014-01-16, see also http://standards.sae.org/j3016_201401/

Challenges

Discuss need for action from an industry perspective	Pave road to market introduction of automated vehicles
Current legal situation does not allow automated vehicles on public roads.	Assess law and identify needed adaption
National laws can be different with respect to automated driving	Analyze main markets - project partners will contribute for their countries
When can a vehicle be considered safe?	Interpretation of existing law. Liability risks?

Research tasks

System classification: Group **categories** of automated driving functions

Legal difficulties for market introduction of automated driving functions:

What are the **new risks** for the manufacturer from product liability	**Usage and protection of data** collected by automated driving functions	**Protection** against corruption and fraud of vehicle data and V2X data

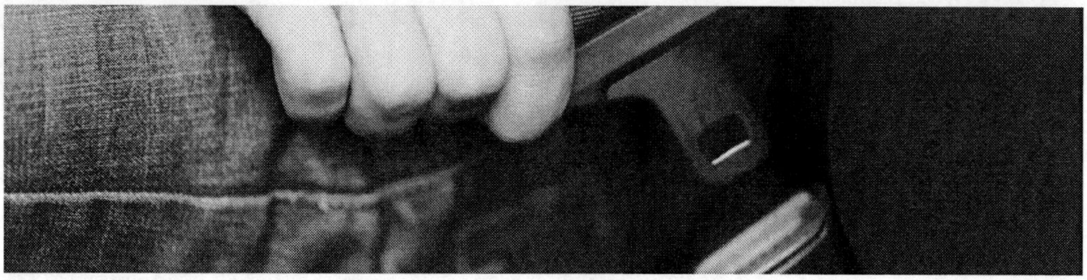

System classification

Systematic derivation of relevant system parameters for:

- Vehicle
- Driver
- Environment

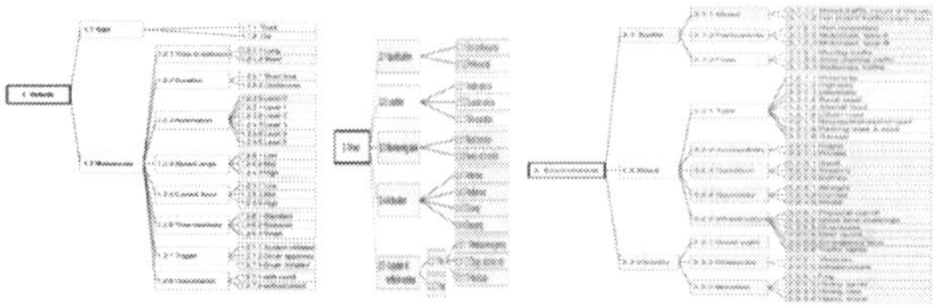

Available to public as Deliverable 2.1 on the AdaptIVe website.

No.	Parameter	Range of values
1	Vehicle type	truck, car
2	Maneuver duration	short, long
3	Maneuver automation	Level 1 - 5
4	Maneuver velocity	low, mid, high
5	Maneuver control force	low, mid, high
6	Maneuver time headway	standard, reduced, small
7	Maneuver trigger	system initiated, driver approved, driver initiated
8	Maneuver Coordination	with coordination, without coordination
9	Driver's location	in vehicle, outside vehicle, tele-operated
10	Road type	type 1 -17 (see Table 4.7)

Scenarios

Representative scenarios for assessment on a case by case basis, needed e.g. for product liability.

Based on system classification.

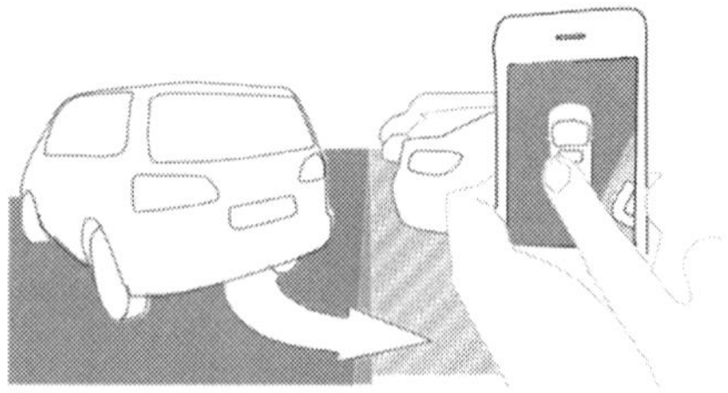

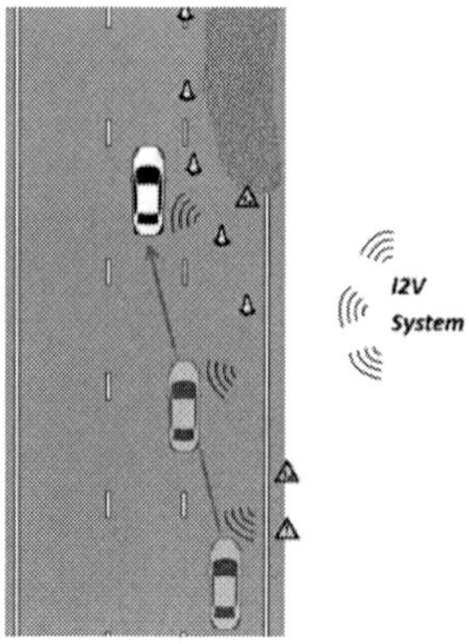

AdaptIVe Workshop on Legal Aspects

- **September 17th, 2015**
- Paris, France
- Hosted at the French Car Manufacturers Assocciation (Comité des Constructeurs Français d'Automobile)

- Discuss legal topics with external stakeholders and expert public
- Promote activities of AdaptIVe
- Get insight in current activities, approaches and status of other projects and working groups

Registration is open and agenda online at www.adaptive-ip.eu

Andreas Knapp

Daimler AG
+49 7031 4389 862
andreas.knapp@daimler.com

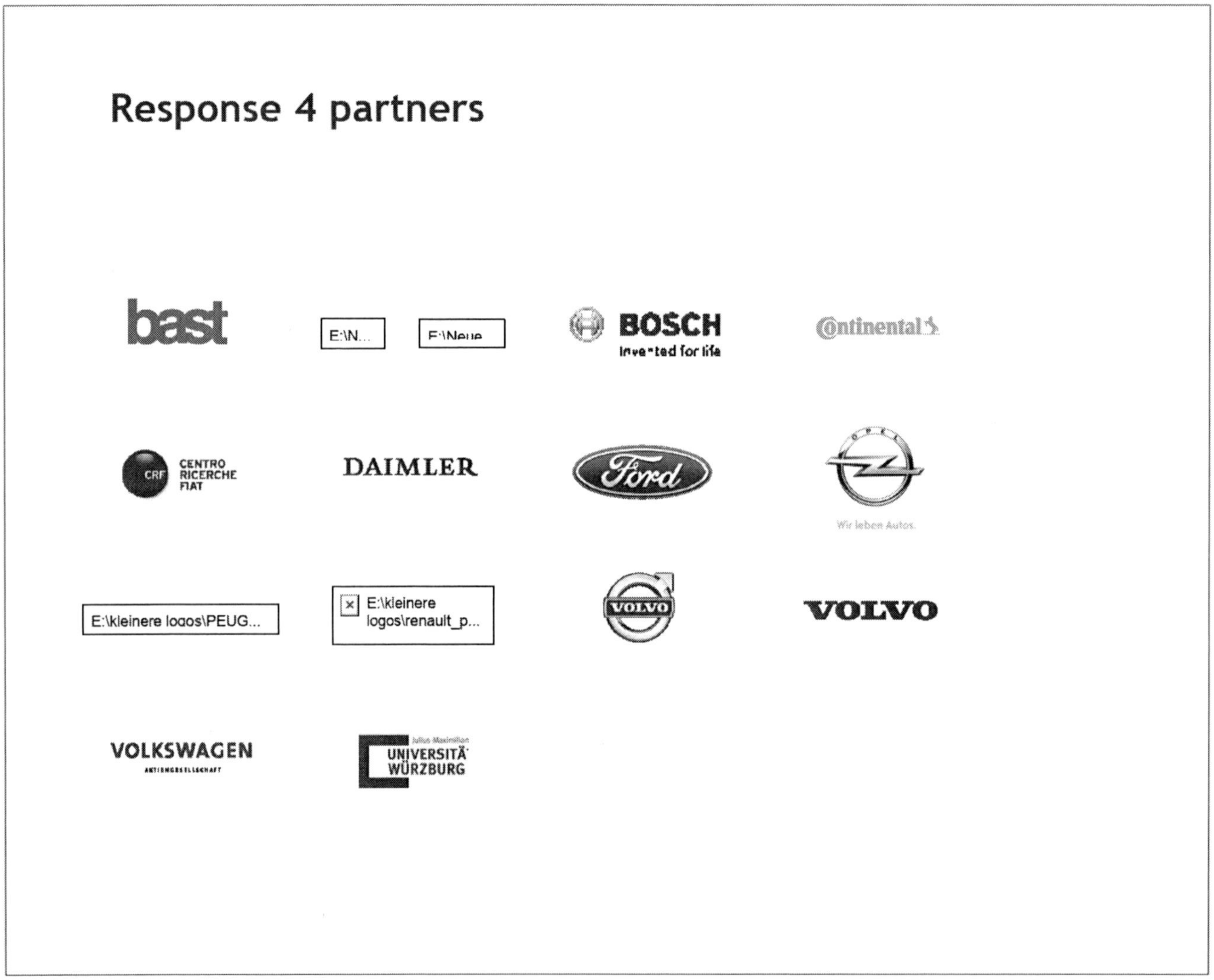
Response 4 partners
bast
E:\N...
E:\Neue
BOSCH
Invented for life
Continental
CENTRO RICERCHE FIAT
CRF
DAIMLER
Ford
OPEL
Wir leben Autos.
E:\kleinere logos\PEUG...
E:\kleinere logos\renault_p...
VOLVO
VOLVO
VOLKSWAGEN
UNIVERSITÄT WÜRZBURG

CAMP
Automated Vehicle Research Consortium
Mercedes-Benz Research & Development North America, Inc. GM TOYOTA
VOLKSWAGEN GROUP OF AMERICA Ford NISSAN
Intelligent Transportation Systems

Development of Safety Principles by automation level

Levasseur Tellis
Ford Motor Company

CAMP AVR Project

- Project Started November 2013
- Project consisted six technical tasks
- Project Participants
  - Ford Motor Company
  - General Motors
  - Mercedes-Benz
  - Nissan
  - Toyota Motors
  - Volkswagen Group of America

Mercedes-Benz Research & Development North America, Inc. GM TOYOTA
VOLKSWAGEN GROUP OF AMERICA Ford NISSAN

AVR Project Objectives

- Develop functional descriptions of automation levels
- Develop list of potential driving automation features
- Develop a set of safety principles that apply by level
- Develop potential objective test methods as a framework for evaluating driving automation systems
- Coordinate with NHTSA
  - Human factors
  - Electronic control systems safety

Why Automation Levels Are Needed

- Critical safety discussions
  - Driver's role changes as automation levels change
  - Proper use of technology
- Common framework
  - Design
  - Customer education/training
  - Regulation
- Benefits to development, understanding and acceptance
  - Categorize technology based on functional attributes
  - Clarify driver's role in proper usage

Development of Top-Level Safety Principles

- A key deliverable of the AVR Consortium entailed
  - The creation of a hazard analysis in order to generate top-level safety principles intended to effectively and succinctly cover the identified hazards inherent in driving automation levels 2-5
  - The development of a set of solution-neutral, top-level, safety principles for each of the driving automation levels defined in a previous AVR task
  - Establish (where possible) safety guidance for driving automation systems, while leaving it to the OEM/system designer to generate plausible solutions

Process to Develop Safety Principles

1. Identify Potential **Losses**
2. Identify Potential **Hazards**
3. Draw **Control Structure**
4. Identify **Undesired Control Actions**
5. Identify **Safety Constraints** (to eliminate undesired control actions)
6. Aggregate constraints into over-arching **principles for each level**

Utilize previous definitions of automation levels throughout

Functional Safety method based on System Theoretic Process Analysis (STPA) developed by Nancy Levesson

Automated Driving Losses and Hazards

Definition of a Loss: *"**An undesired and unplanned event that causes human injury or property damage.**"*

Definition of a Hazard: **"A system state that together with a worst-case set of external disturbances may lead to a loss."**

Loss	**Vehicle Collision with a Threatening Object**
H1	**Vehicle leaves the roadway**
H2	**Vehicle loses traction or stability**
H3	**Vehicle comes too close to threatening objects in the roadway**
H4	**Vehicle violates traffic laws, rules, and norms**

Note: By definition, hazards are identical for all levels of automation

Task 6 – Safety Principles

Three Actors Engaged in Driving Automation

Vehicle Operator*

Driving Automation

Vehicle Systems

All three are necessary to describe how automation impacts the performance of the dynamic driving task (DDT)

* - *e.g., driver*

Clear roles of the driver and automation critical in development of Safety Principles

Dynamic Driving Task (DDT) Allocation:

- DDT must be entirely completed on a sustained basis
- Driver performs all aspects of DDT not performed by the driving automation system

Safety outcome of automated driving:

- The potential of automated driving can only be realized if the driver understands their role in DDT completion
- Driver needs to complete remaining portions of DDT unless complete replacement by the driving automation system is available
- Driver's use of the driving automation system plays a significant role in net safety benefit

Complete Dynamic Driving Task

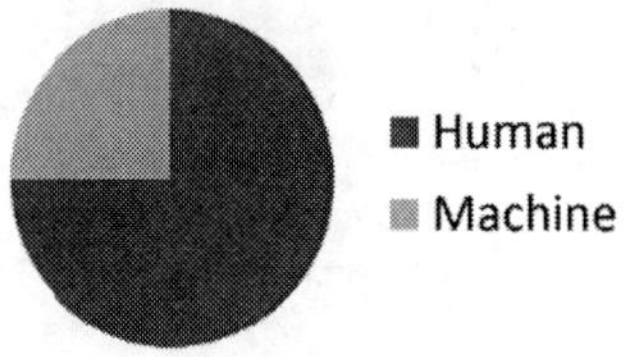

Incomplete Dynamic Driving Task

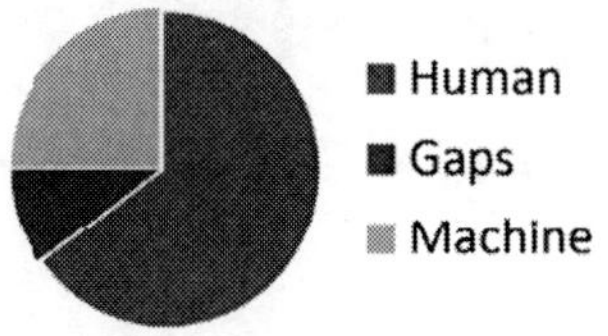

Safety Constraints Aggregated to Safety Principles

Undesired Control Action	Safety Constraint	Potential Principles
Human driver provides incorrect control when control not needed, when human OEDR is not engaged due to lack of attention or understanding.	Human OEDR shall be engaged	For Level 2 automation, avoidance of hazards is dependent on the human driver performing the OEDR subtask and completing the DDT.
Human driver provides incorrect control when control not needed, when human OEDR is not engaged due to inability to perceive environment.	Human OEDR shall be engaged; human driver shall able to perceive environment.	For Level 2 automation, avoidance of hazards is dependent on the human driver performing the OEDR subtask and completing the DDT. Human driver must be able to perceive environment.
Human driver does not provide correct control, or provides control incorrectly, late or early, when control is needed to avoid a hazard, when automation is inactive.	Human driver shall correctly control the vehicle when control is needed to avoid a hazard and automation is inactive	For Level 2 automation, avoidance of hazards is dependent on the human driver performing the OEDR subtask and completing the DDT.

Principle on human driver OEDR

Principle on vehicle design

Principle on human driver OEDR

Level 2 Principles

- Driver
  - Operational Readiness
  - OEDR
  - Decision to initiate/ override cancel automation
  - Fallback control in the event of vehicle or automation failure
- Vehicle
  - Vehicle Controls
  - Visibility
  - Driver control
- Automation - controllability

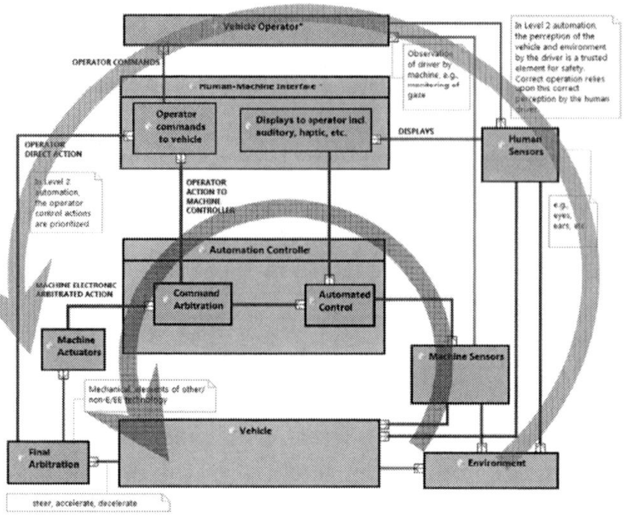

Level 3 Principles

- Driver
  - Operational Readiness
  - Decision to initiate/ override/ cancel automation
  - Fallback control in event of vehicle failure
- Vehicle
  - Vehicle Controls
  - Visibility
  - Driver Control
- Automation
  - Driver initiated
  - Persistent indication of high automation
  - Complete OEDR
  - Validate operational domain
  - Controllability during override or cancel

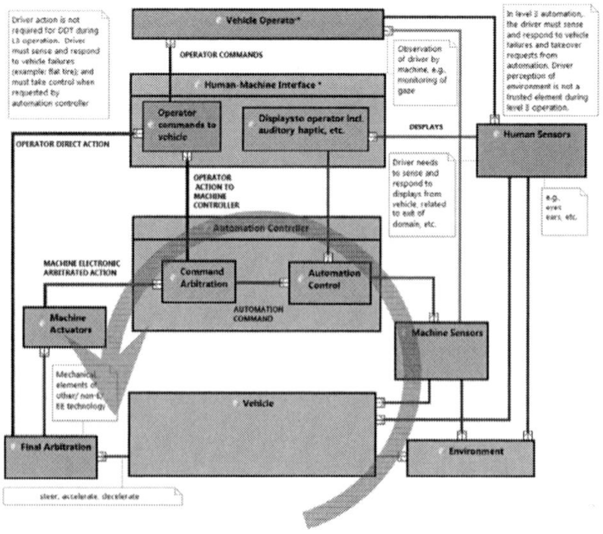

Level 4 & 5 Principles

- **Driver/Operator**
  - Operational Readiness
  - Decision to initiate operation
- **Vehicle**
  - Driver controls if low automation available
  - Visibility if human driver is present
- **Automation**
  - Complete OEDR
  - Persistent indication of high automation if human driver is present
  - Fail safe operation
  - Validate operational domain
  - May not immediately respond to driver request

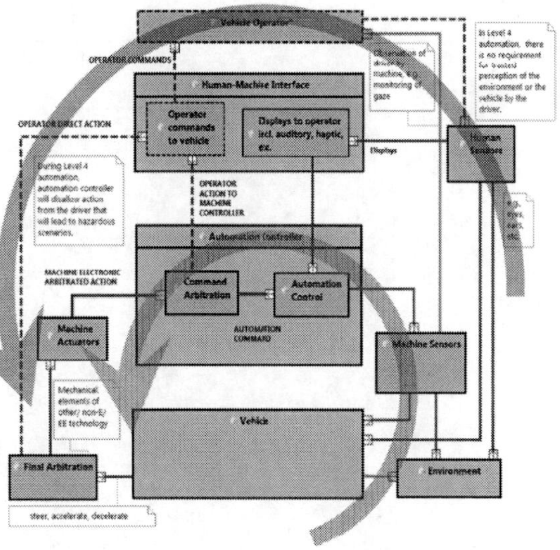

Summary of Principles – I

Safety principle related to:	Level 2	Level 3	Level 4	Level 5
	When automation is engaged at:			
Driver/ Operator	assures operational readiness (SP 2.1 i)	← (SP 3.1)	← (SP 4.1)	← (SP 5.1)
	relied upon to avoid hazards, by completing the OEDR subtask and DDT (SP 2.1 ii)			
		activates automation for first time in drive cycle (SP 3.4)	← (SP 4.5)	← (SP 5.5)
		determines if vehicle failure occurs and takes control (SP 3.10 i)		
		understands that direct driver input will cause a transition to lower lever automation and driver will then control those inputs (SP 3.10 ii)		
		takes control when requested by automation (SP 3.10 iii)		
		understands that after automation request to take control, automation will only remain in control for a limited time (SP 3.10 iv)		
Vehicle systems	designed such that the driver is capable of fully performing DDT (lateral / longitudinal control and OEDR) (SP 2.2)	← (SP 3.2)	← Note - include if vehicle is capable of lower level automation	← Note - include if vehicle is capable of lower level automation

Summary of Principles – II

Safety principle related to:	When automation is engaged at: Level 2	Level 3	Level 4	Level 5
Automation Controller (part 1)	arbitrate between defined driver inputs and driving automation commands by prioritizing the driver input (SP 2.3 i)	← (SP 3.3 i)		
	allow driver to take full control at any time (SP 2.3 ii)	← (SP 3.3 ii)		
	may verify defined driver input before deactivating driving automation (SP 2.3 iii)	← (SP 3.3 iii)		
		provides persistent indication to driver of operation in high automation state (SP 3.5)	← Note - include if vehicle is capable of lower level automation	← Note - include if vehicle is capable of lower level automation
		provides indication to driver of request to transition to lower level automation (SP 3.6)	← Note - include if vehicle is capable of lower level automation	← Note - include if vehicle is capable of lower level automation
		competently performs the DDT within its operational design domain (SP 3.7 i)	← (SP 4.2 i)	competently performs the DDT in all domains (SP 5.2)

Summary of Principles – III

Safety principle related to:	When automation is engaged at: Level 2	Level 3	Level 4	Level 5
Automation Controller (part 2)		prohibit entry into automated driving when domain is not achieved (SP 3.7 ii)	← (SP 4.2 ii)	
		vehicle/automation system single point failure shall not cause immediate loss of total control (SP 3.8)	designed such that any single failure does not lead to a hazardous situation (SP 4.3)	← (SP 5.3)
		before exiting domain or in advance of automation failure that impacts DDT performance, system shall transfer control to the driver (SP 3.9)	ability to engage minimal risk condition when necessary (SP 4.2 iii)	← (SP 5.2)
		verified driver inputs shall cause transition to lower level automation (SP 3.9 i) system shall maintain operational condition that affords reasonable transition time to driver (SP 3.9 ii)	may delay requests by operator to take over/stop automation when necessary to avoid identified hazards (SP 4.4)	← (SP 5.4)

Key Take-aways

- Hazards and losses to be considered in the development of safety principles per driving automation level were developed
- Generic control structure was put together to describe each driving automation level and facilitate the creation of safety principles by level
- Level 2 driving automation systems are intended to complement but not substitute for the human driver in performing the Dynamic Driving Task (DDT) and the safety principles developed for Level 2 highlight this intention.
- The most significant safety principle placed on a Level 3 driving automation systems is that when it is engaged, the DDT is performed solely by the driving automation system within a limited ODD (e.g., geographical location, environmental condition, speed, etc.
- The most significant safety principle placed on Level 4 and 5 driving automation systems is that operator requests to take over part or all of the DDT may not be immediately granted as the system may be operating in domains where human control is not allowed or could cause an undesired hazard/control action.

BACKUP

Lower Levels (1-2) of Automation

Automation Level Name	Automation Level Narrative Description	Dynamic Driving Sub-Tasks		Functional Capability	
		Sustained Execution of Lateral and/or Longitudinal Control	Object & Event Detection and Response (OEDR)	Fallback Performance of Dynamic Driving Task	Driving Mode Circumstance, Location Capabilities
Driver performs all or part of the dynamic driving task and general system functional capabilities					
0 No Automation	The full-time performance by the driver of all aspects of the dynamic driving task, even when enhanced by warning or event-based intervention systems	Driver	Driver	Driver	None of the DDT is automated
1 Driver Assistance	The driving mode-specific execution by a system of either sustained lateral **OR** sustained longitudinal control using sensing data and with the expectation that the driver performs the remainder of the dynamic driving task	Driver and system	Driver	Driver	Some driving modes
2 Partial Automation	The driving mode-specific execution by one or more systems of both sustained lateral **AND** sustained longitudinal control using sensing data with the expectation that the driver performs the remainder of the dynamic driving task	**System**	Driver	Driver	Some driving modes

Higher Levels (3-5) of Automation

Automation Level Name	Automation Level Narrative Description	Dynamic Driving Sub-Tasks		Functional Capability	
		Sustained Execution of Lat. and/or Long. Control	Object & Event Detection and Response (OEDR)	Fallback Performance of Dynamic Driving Task	Driving Mode Circumstance, Location Capabilities
Automated driving system performs all the dynamic driving task and general system functional capabilities					
3 Conditional Automation	Automated driving system features in this level automate the complete dynamic driving task, providing appropriate responses to relevant objects and events. However, the automation is situationally-limited in functional capabilities both in terms of driving modes, circumstances, and/or locations and, particularly in terms of fallback performance capability. There is the expectation that the driver will respond appropriately to a request to resume performance of the dynamic driving task	System	**System**	Driver	Some driving modes
4 High Automation	The driving mode-specific performance by a system of all aspects of the dynamic driving task, providing appropriate responses to relevant objects and events, even if a driver does not respond appropriately to a request to resume performance of the dynamic driving task	System	System	**System**	Some driving modes
5 Full Automation	The full-time performance by a system of all aspects of the dynamic driving task, providing appropriate responses to relevant objects and events, under all on-road driving conditions legally available to a driver	System	System	System	
Automated driving system performs complete dynamic driving task, providing appropriate responses to relevant objects and events, and greater functional capability					

The UM Mobility Transformation Center

Research Roadmap

July 2015

John Maddox — Assistant Director
Peter Sweatman — Director
Huei Peng — Associate Director
Jim Sayer — Deployment Director

The societal costs of transportation

6 Million+ Crashes every year
32,000+ deaths annually

Leading cause of death for ages <34
Estimated 1 - 2 million deaths globally

$240B death and injury
$100B stuck in traffic

30 % of carbon emissions
70 % of oil consumption

US Trends & Assumptions

- Population growth
  - Additional ~100M Americans by 2050
- Transportation will continue to be a major enabler for the US economy
- VMT will increase 50% by 2050 to 4.5T
- Interstates/Highways will still carry largest portion of travel
- Energy costs will increase
- Congestion will become much worse for many cities and regions – usage restrictions
- New business models will evolve (car sharing, VMT-based usage fees, etc)

- These trends, if not addressed, could cause erosion in
  - quality of life
  - economic growth
  - national economic competitiveness

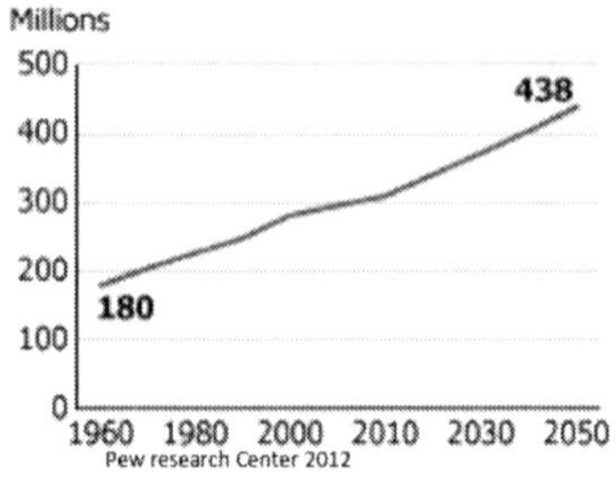

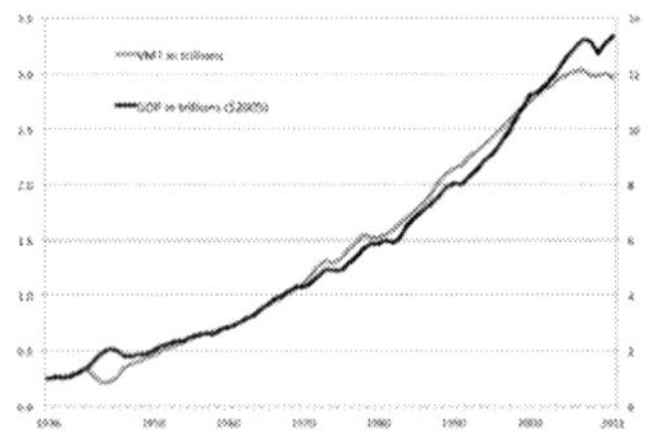

The MTC

A public/private R&D partnership
that will lead a revolution in mobility
and develop the foundations
for a commercially viable ecosystem
of connected and automated vehicles

Prototyping an entire system of connected and automated transportation on the streets of southeastern Michigan through 2021

a shared initial investment of $100M over 8 years with 25% coming directly from the U-M

MTC Roadmap Process

- Identify technology opportunities
  - To address key problems
  - To provide new business models
- Strategize with partners
  - Multiple industries
  - Multiple government agencies
- Identify critical barriers
  - Mostly public-private in nature
- Identify key research thrusts, questions and technology needs
  - Prioritize and conduct research and development projects
  - Many will be interrelated and iterative
- Create deployments to prototype a new transportation system
  - Overcome critical barriers
- Quantify and understand outcomes

Technology Applications

	applications
Connected	
V2V	Cooperative Crash Imminent Braking
	Emergency Brake Lights
	Intersection Movement Assist
	Left Turn Assist
	Blind Spot / Lane Change Warning
	Do Not Pass Warning
	Right Turn in Front of Transit Vehicle, etc
+ V2I	Adaptive Signal Control
	Advanced Traveller Info System
	Work/Speed Zone warning
	Motorist Weather Advisories/Warnings
	Pedestrian in Crosswalk
	Agency Probe Data
	Smart Parking, etc
Automated	
AV	L2 lane keeping
	L2 traffic jam assist
	L3 highway autopilot?
	L2/3 parking assistant
Automated + Connected	
AV + V2V + V	all above, in more scenarios, +
	L3 platooning
	L4 platooning
	L2/3/4 auto intersection eco-approach/departure
	L4 auto-valet
all + V2P	L4 full urban operation amongst pedestrians

Technology Pros & Cons

	pros	cons
Connected		
V2V	proven effective for safety - avoiding collisions	relies on driver reaction
	inexpensive - can be applied on many vehicles	relies on other vehicles (critical mass)
	retrofittable	requires security system
	sees around corners	perception of privacy loss
	sees through fog/rain/snow	
	sees at longest sensor range	
	sees multiple vehicles ahead	
	knows much more about road partners	
+ V2I	enhances mobility thru adpative signal control	requires infrastructure investment
	enhances energy-use through eco routing/timing	requires added security system
	enables weather apps	further perception of privacy loss
	enables 0/1st level eco-driving	
	enables pedestrian detection in crosswalk	
	enables smart parking	
Automated		
AV	relies less on human intervention	expensive - can't be applied on many vehicles
	doesn't rely on other vehicles	technology not fully ready
	added driver convenience	not yet proven reliability
	high customer interest	not retrofitatable
	potential for improved safety, unproven	requires security system
	potential for improved eco-driving	requires policy decisions/changes
		unclear if all veh data can remain private/anonymous
		may add VMT due to convenience

Technology Pros & Cons

Automated + Connected		
AV + V2V + V2I	all advantages above	most disadvantages above
	adds reliability to sensing & decision making	added cost
	enables platooning at close following distance	added proveout requirements
	enables safer lane changing and passing	requires comprehensive security system
	communicates locations of map changes/updates	
	communicates road construction / maintainence	
	enables 2/3/4th level eco-driving	
	enables L4 driverless (on most roads)	
all + V2P	enables safe urban operation around pedestrians	requires smart phone/device solution
	enables rapid retrofit system	

Relative Benefits (1-10 relative scale)

		Safety	Mobility	Environment	Convenience
Connected		?	?	?	
V2V		5	2	2	0
V2I		1	5	3	1
Automated		?	?	?	?
L1		2	0	1	1
L2		0	0	0	1
L3		1	0	2	3
L4		1	3	2	4
Con+Auto		10	10	10	10

Relative Benefits (1-10 relative scale)

		Safety	Mobility	Environment	Convenience
Connected		?	?	?	
V2V		5	2	2	0
V2I		1	5	3	1
Automated		?	?	?	?
L1		2	0	1	1
L2		0	0	0	1
L3		1	0	2	3
L4		1	3	2	4
Con+Auto		10	10	10	10

Relative Benefits (1-10 relative scale)

		Safety	Mobility	Environment	Convenience
Connected		?	?	?	
V2V		5	2	2	0
V2I		1	5	3	1
Automated		?	?	?	?
L1		2	0	1	1
L2		0	0	0	1
L3		1	0	2	3
L4		1	3	2	4
Con+Auto		10	10	10	10

Relative Benefits (1-10 relative scale)

		Safety	Mobility	Environment	Convenience
Connected		?	?	?	
V2V		5	2	2	0
V2I		1	5	3	1
Automated		?	?	?	?
L1		2	0	1	1
L2		0	0	0	1
L3		1	0	2	3
L4		1	3	2	4
Con+Auto		10	10	10	10

Relative Benefits (1-10 relative scale)

		Safety	Mobility	Environment	Convenience
Connected		?	?	?	
V2V		5	2	2	0
V2I		1	5	3	1
Automated		?	?	?	?
L1		2	0	1	1
L2		0	0	0	1
L3		1	0	2	3
L4		1	3	2	4
Con+Auto		10	10	10	10

Relative Benefits (1-10 relative scale)

		Safety	Mobility	Environment	Convenience
Connected		⍰	⍰	⍰	
V2V		5	2	2	0
V2I		1	5	3	1
Automated		⍰	⍰	⍰	⍰
L1		2	0	1	1
L2		0	0	0	1
L3		1	0	2	3
L4		1	3	2	4
Con+Auto		10	10	10	10

Need Connect-imated !

Partners

Industry

- Auto manufacturers
- Components and systems
- Telecommunications
- Big data management
- Freight
- Traffic control systems
- Insurance
- Public transportation
- Payment systems
- Smart parking

Government

- USDOT
- MDOT
- USDOE
- MEDC-Automotive Office
- City of Ann Arbor
- Other federal, state, and local agencies

Academia

- U-M
- TTI
- U-Md CATT Lab
- Other leaders

MTC AFFILIATES

- Auto Club Enterprises an AAA affiliate
- AGC Automotive
- Allstate Insurance Co.
- Arada Systems, Inc.
- Autoliv
- Brandmotion LLC
- Calspan Corporation
- Changan Automobile
- Cohda Wireless
- Desjardins
- DURA Automotive Systems
- Faurecia
- Harada Industry of America, Inc.
- Harman International Industries
- HERE, a Nokia company
- Hitachi, Ltd.
- IAV
- IDIADA
- LG Electronics
- Munich Re
- Nexteer Automotive
- OSIsoft, LLC
- PTC, Inc.
- Realtime Technologies, Inc.
- Renesas Electronics America Inc.
- Savari Inc.
- Subaru
- Sumitomo Electric Industries, Ltd.
- Suncorp Group
- TASS International, Inc.
- TRW Automotive
- Zip Car

UNIVERSITY PARTNERS

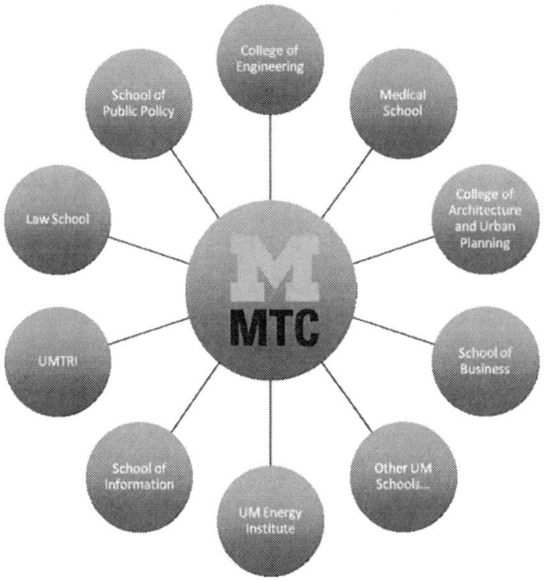

CRITICAL BARRIERS: AUTOMATION

- Legal, Liability and Insurance issues
- Customer acceptance & understanding
- Standardized technology assessments & validation for safe operation
- AV operation in mixed traffic
- Catalog of human behavior in critical situations
- Transition of vehicle control with partial automation
- Cybersecurity of AVs

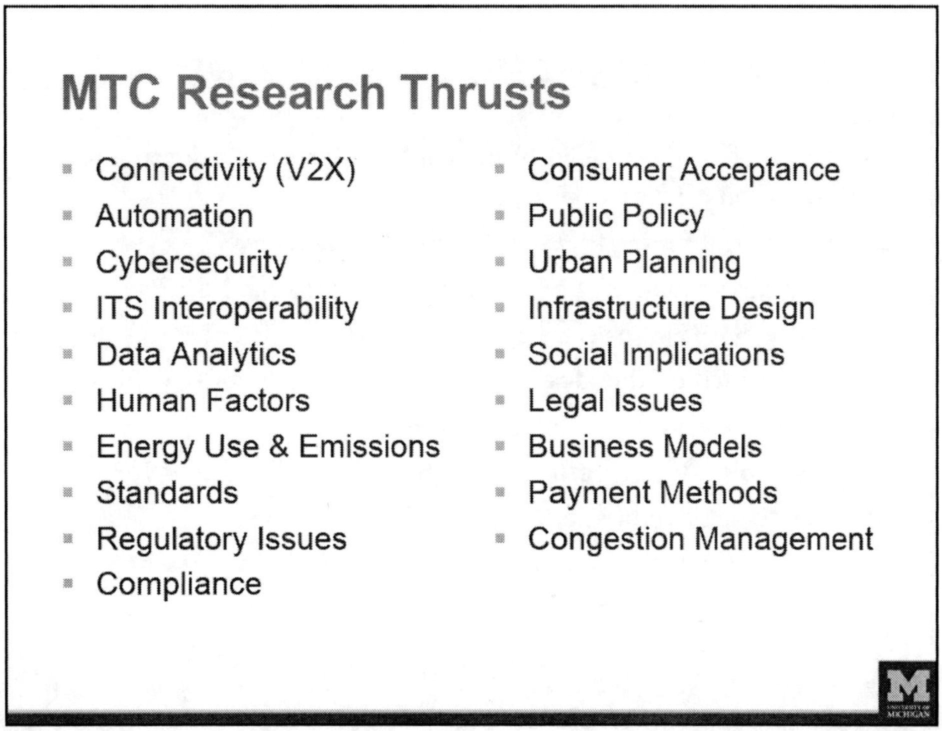
MTC Research Thrusts
Connectivity (V2X)
Automation
Cybersecurity
ITS Interoperability
Data Analytics
Human Factors
Energy Use & Emissions
Standards
Regulatory Issues
Compliance
Consumer Acceptance
Public Policy
Urban Planning
Infrastructure Design
Social Implications
Legal Issues
Business Models
Payment Methods
Congestion Management

Prioritized MTC Research Thrusts
Connectivity (V2X)
Automation
Cybersecurity
ITS Interoperability
Data Analytics
Human Factors
Energy Use & Emissions
Standards
Regulatory Issues
Compliance
Consumer Acceptance
Public Policy
Urban Planning
Infrastructure Design
Social Implications
Legal Issues
Business Models
Payment Methods
Congestion Management

PRIORITY RESEARCH QUESTIONS

- What role do test facilities have in standardized assessment & validation for safe operation?
- How much is enough when testing for reliability?
- How can simulation be leveraged to assess readiness?
- How will AVs interact with human-driven vehicles?
- What specific value does V2X bring to an automated vehicle?
- What role does the built infrastructure have?
- What roles do the data and mapping infrastructure have?
- How to achieve public acceptance?
- Can we determine a benefit to non-connected and/or non-automated vehicles in the environment?

CROSS CUTTING RESEARCH QUESTIONS

- Cybersecurity
  - What are the specific needs/risks of AV's?
  - How do you provide secure over the air updates?
- Legal/Liability/Insurance
  - What changes might be required of the existing legal system?
  - .... of the existing insurance system?
  - How will fault be assessed in AV crashes?
  - What are the key privacy impacts?
- Customer Value
  - How do you define and measure customer value and acceptance for all stakeholders?

CROSS CUTTING RESEARCH QUESTIONS

- Data Collection/Analytics
  - What data should be stored on an AV to aid in determination of root cause of crashes/malfunctions?
  - What existing data sets could be leveraged with new CAV data sets?
  - What deployment data should be collected and what are the uses?
  - How can data drive entrepreneurship and new business models?
- Societal Impacts
  - What are the impacts on energy and health?

RESEARCH APPROACH

- Industry Working Groups
  - Subject matter experts come together to direct activities
    - Research activities
    - Pillar deployment activities
    - Outreach and engagement activities
  - Co-chaired by Industry and UM-MTC experts
- Current active Working Groups
  - Pillar 1
  - Pillar 2
  - Pillar 3 & Mcity
  - Cybersecurity
  - Legal and Insurance

RESEARCH APPROACH

- Faculty RFP
  - UM Faculty develop proposals to address priorities
  - Research and "pillar-development" projects
  - 1st round funded June 2104
  - 2nd round funded May 2015
  - Approximately $3.1M to-date
  - Status and results presented to MTC LC and Affiliate members, and government agencies
  - MTC working groups prioritize next needs
- MTC will invite faculty from other universities to participate

RESEARCH PROJECTS

1st Round	
RoadMap for Automated Vehicles in Ann Arbor	School of Urban Planning
Cybersecurity Testing Methodologies and Tools for CAVs	UMTRI, COE
The Driver in the "Driverless" Car: Developing a Simulator Approach for Evaluation of Driver Performance and Behavior	UMTRI
Mobility Data Mining for Intelligent Transportation	UMTRI, School of Information
2nd Round	
Cybersecurity Roadmap for AVs	UMTRI, COE
Application of Current Legal Precedents on Fault and Liability for Crashes of Automated Vehicles	UMTRI, School of Public Health
Examination of Driver State Monitoring and Operator Engagement as Strategies for Mitigating Human Factors Challenges with Transfer-of-Control During Automated Driving	UMTRI
Survey and Framework for Legal and Regulatory Issues Arising from AV Technologies	Law School
Remote Intrusion Detection and Prevention for CAV Cybersecurity	COE
Investigation of Drivers Adaption Behavior and Decision Making when Interacting with CAV Technologies	UMTRI
Consumers response to AVs	UMTRI
Age-Related Differences in Driver behavior with AVs and Transfer-of-Control	UMTRI/ATLAS
Intelligent Parking Guidance System for CAVs based on Connected Sensor Networks	COE/UMTRI
Development of "Smart Carts" as a Shared AV Transportation System	COE
Development of Pedestrian-to-Vehicle Communication for CAV Deployments	UMTRI
Ethics and Automation – A Sudden-Reveal Study	UMTRI

Deployments are necessary to address all of those research areas in a comprehensive and accelerated manner

They will become living laboratories for public and private efforts

MTC Platforms for Innovation

Three pillar programs, in collaboration with MDOT:

1. **Ann Arbor Connected Vehicle Test Environment (2014+)**
2. **Southeast Michigan Connected Vehicle Deployment (2015+)**
3. **Ann Arbor Automated Vehicle Field Operational Test (2016+)**

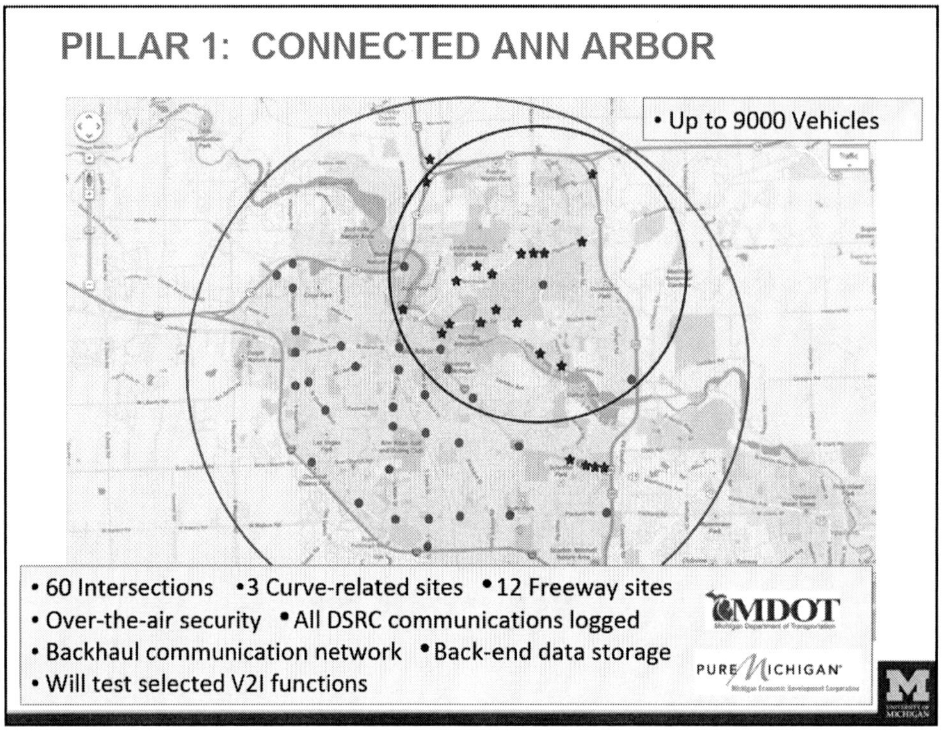
PILLAR 1: CONNECTED ANN ARBOR
• Up to 9000 Vehicles
• 60 Intersections •3 Curve-related sites •12 Freeway sites
• Over-the-air security •All DSRC communications logged
• Backhaul communication network •Back-end data storage
• Will test selected V2I functions
MDOT
PURE MICHIGAN

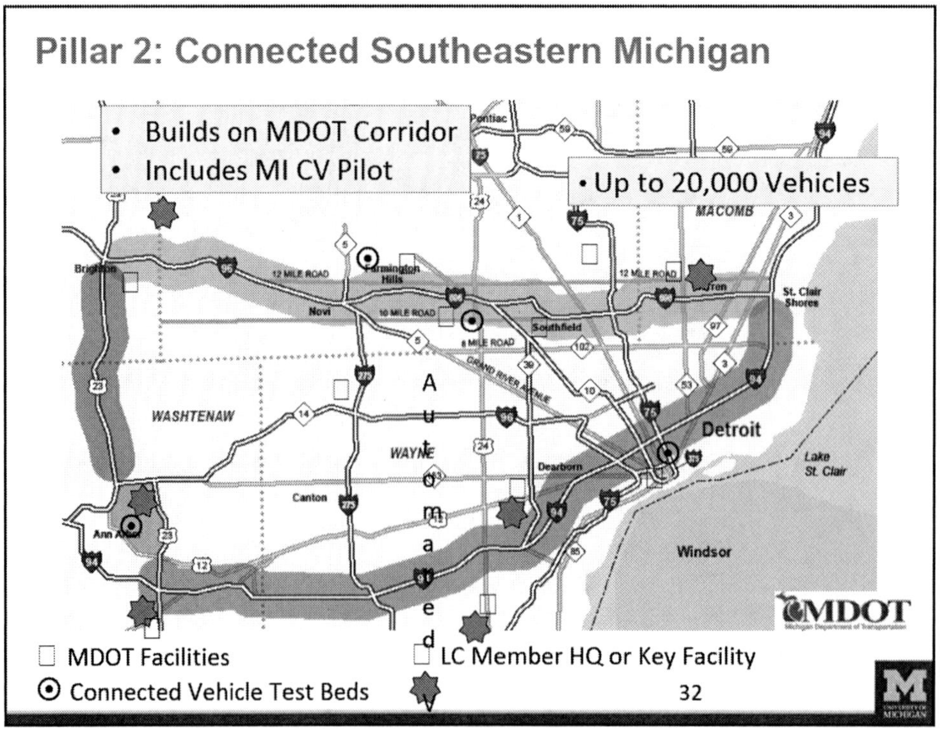
Pillar 2: Connected Southeastern Michigan
• Builds on MDOT Corridor
• Includes MI CV Pilot
• Up to 20,000 Vehicles
Farmington Hills
Novi
Southfield
Brighton
WASHTENAW
WAYNE
MACOMB
Canton
Ann Arbor
Dearborn
Detroit
St. Clair Shores
Lake St. Clair
Windsor
12 MILE ROAD
10 MILE ROAD
8 MILE ROAD
GRAND RIVER AVENUE
Automated
MDOT Facilities
Connected Vehicle Test Beds
LC Member HQ or Key Facility
MDOT
32

PILLAR 3: AUTOMATED ANN ARBOR

Ann Arbor Automated Vehicle Field Operational Test (2016+)

- 2,000 connected and automated vehicles
- Including Level 4 automated vehicles
- 27 sq. miles of densely instrumented infrastructure

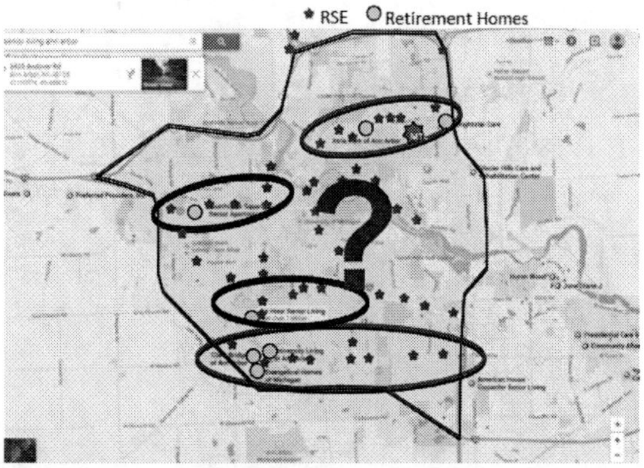

PILLAR 3: Mcity

- Safe, repeatable, off-roadway test environment for AVs: simulated city
- Technology research, development, testing, and teaching
  - $6.5M project; $3.0M funding from MDOT
  - Construction commenced July 15 2014
  - Grand opening: July 20, 2015

MDOT

2015 Calendar

NECTAR Conference – Urban Planning	June 14-16
Mcity Grand Opening	July 20
TRB/AUVSI Automated Vehicle Symposium	July 21-23
Academic Partner Meeting	July 24
LC Strategy Meeting & Research Review	Sep 21
MichAuto Summit & MTC Congress	Sep 30
UMTRI Transportation Safety Symposium	Oct 15
MTC Research Review - General Session	Dec 15
Potential next round RFP	Dec/Jan

Ethical Considerations for Vehicle Automation Systems

J. Christian Gerdes

Professor of Mechanical Engineering and, by courtesy, of Aeronautics and Astronautics

Director, Revs Program at Stanford and Center for Automotive Research at Stanford

REVS program at STANFORD

Workshop at Stanford University, June 5 2015

Daimler und Benz Stiftung

Ethics Autonomous Vehicles

Professor Patrick Lin
Philosophy
California Polytechnic University

REVS program at STANFORD

Ethics is a process, not a module

Professor Shannon Vallor
Philosophy
Santa Clara University

Professor John Sullins
Philosophy
Sonoma State University

Are we programming ethical decisions or ethically programming?

REVS program at STANFORD

SAFETY
LEGALITY
MOBILITY

Break the law?
Wait all day?

1st Person versus 3rd Person Ethics

Professor Debra Satz
Philosophy
Stanford University

Professor Erin MacDonald
Mechanical Engineering
Stanford University

REVS

Traffic is a social dance

Lessons from other disciplines

Professor Mykel Kochenderfer
Aeronautics and Astronautics
Stanford University

Dr. Noah Goodall
Virginia Center for Transportation
Innovation and Research

Philosophers and Engineers Working Together!

- Map philosophical concepts to programming constructs
  - Deontological ethics to constraints
  - Consequentialism to cost functions
- Can balance objectives using a model predictive control framework
  - But what weights do we use?

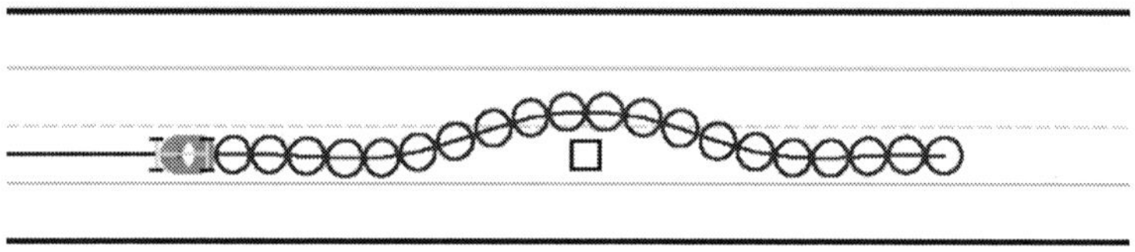

- Moral laws modeled analogously to physical laws
  - A parallel suggested by Immanuel Kant

REVS

Where Can I Find More?

Chapters by conference alumni including:

Eva Fraedrich
Thomas Gasser
Chris Gerdes
Barbara Lenz
Patrick Lin
Bryant Walker Smith
Stephen Wu

About 25% in English, full translation coming

Just Search 'Springer Autonomes Fahren'
Free through Open Access!

REVS

REVS

program at STANFORD

REVS *program at* STANFORD

13

TRB

Towards Road Transport Automation
Opportunities in Public–Private Collaboration
Third EU–U.S. Transportation Research Symposium

National Academy of Sciences Building
Washington D.C.
Informal report

Peter Sweatman (Chair)
Maxime Flament (Co-Chair)
Bob Denaro

Outline

- Symposium concept
- Event April 14-15 2015
  - NAS Washington DC
- Beyond the technology
  - Economic, environmental and societal implications
- Identifying opportunities for research collaboration

Symposium Concept

Planning Committee

US

- Peter Sweatman, University of Michigan Transportation Research Institute, *Chair*
- David Agnew, Continental Automotive
- Robert Denaro, ITS Consulting
- Ginger Goodin, Texas A&M Transportation Institute

EU

- Maxime Flament, ERTICO–ITS Europe, *Vice Chair*
- Roberto Arditi, SINA Group
- Aria Etemad, Volkswagen AG, Germany
- Natasha Merat, University of Leeds

4

Towards Road Transport Automation:

Opportunities in Public-Private Collaboration

5

Mission

What are the **complementary roles and responsibilities** of the actors in a **Public-Private ecosystem** needed to drive the **evolution of the automated vehicles** towards a 21th century mobility system **(integrating and optimising vehicle, user, and infrastructure)**?

6

Expected Outcome

- Foster Transatlantic Partnerships and future collaboration on research areas of mutual interest
- Draw out research challenges worthy of international collaboration

7

Key Elements

White papers, Constituencies, Key Topics, Use Cases

White Paper #1
Road Transport Automation as a Public–Private Enterprise, **Steven Shladover and Richard Bishop**

- Diversity of automation concepts
- Diversity of operational environments
- Different deployment approaches: everything somewhere vs something everywhere
- Need for support from infrastructure
- New business models emerging
- How safe is safe enough?

White Paper #2
Road Transport Automation as a Societal Change Agent, **Risto Kulmala and Oliver Carsten**

- Significant potential benefits both in short and long term but disadvantages exist as well
- High uncertainties on best deployment models
- Major challenges related to mixed traffic and other vulnerable road users
- Potentially higher costs of operation allocated to all road transport actors e.g.training, maintenance, periodic inspections, signage, road markings, digital infrastructure, accurate, traffic information

Constituencies

- Automotive (8)
- Authorities (5)
- Infra/Road Operators (6)
- Public Transport (3)
- Goods Transport (3)
- Users/Drivers/VRU (2)
- Shared Vehicles/Fleet (1)
- Insurers (2)
- Service Providers (4)
- Research (12)
- Legal/Lawyers (2)

11

Key Topics

- Technology
- Legal
- Business Models
- Human Factors
- Security
- Policy Making
- Testing
- Acceptance

12

Use Case Scenarios

Use Case 1
Moderately Automated Highway Operation (Platooning)

Use Case 2
Highly Automated Urban Operation

Use Case 3
Fully Automated Tailored Mobility Service (Urban Chauffeur)

Automated Driving Use Cases

USE CASE	Level of Autom. (SAE)	Speed	Dedicated space needed	Private or public	Examples (projects)
1 Freeway platooning	2-3	High (> 70 mph)	Possibly both	BOTH	Sartre, Peloton
2 Automated city centre	3-4	Low (10-40 mph)	No	PRIVATE	Adaptive
3 Urban Chauffeur	4	Low (< 25 mph)	Both	PUBLIC	Google, Citymobil2

Preliminary Observations

a personal sampling

Use Case #1: Freeway platooning

Moderately automated highway operation

- May be a good business case for fleets but it addresses only highways and limited transport issues
  - What is the benefit across fleets? Who should be first in line? How does it affect non-users?
  - Early benefits modest due to wider gaps and slower speeds for vehicles
- Challenges related to functional safety, dumb trailers, acceleration and braking capacity, cooling vents, lead driver responsibilities, planning of platoons, fleet relations, acceptance
- Liability, driver training and licensing issues may be overcome

Large-scale platoon pilots and field tests are needed for further learnings

Use Case #2: Automated City Center
Highly automated urban operation, low-speed, no dedicated space

- Focus on improving safety and efficiency where it is most needed i.e. in urban environment
- Opens for sharing economy solutions
- May not answer some of the current trends in urban development policies
- Challenges in human factors in mixed urban traffic, urban traffic management, needs for investment in facilities, certification of roads and vehicles, liability of L3, quantification of impacts and costs, business models, role of collected data, need for AI & machine learning.

Most urgent research needs:
Human Factors, Legal and liability framework and evaluation of impacts

Use Case #3: Urban Chauffeur
Highly automated urban mobility service, low-speed, dedicated or shared space

- Offers large savings for first- last- mile of public mass transit, transport accessibility and urban goods deliveries
- Opens to new urban center design in-line with "liveable cities" concept
- Reduces the need and usage of private cars
- Requires political courage and careful community consultations: regulatory barriers
- Not clear public acceptance for trading off status quo
- Cybersecurity and data privacy concerns
- High safety paramount, certification issue
  - How safe is safe enough?
  - Interactions with vulnerable road users

Most urgent research needs: Large scale trials, interaction with VRU, minimum standards and performance requirements, impact, acceptance, cybersecurity, certification models

Other takeaways

- Never underestimate the power of status quo
- Importance of data collection, sharing and analysis is underestimated
- Level 3 may not be viable from a liability stand point
- Levels of automation are helping the expert communities but are not designed for the wide public: functionalities will be the end-products
- Keep the end-users in mind, solving problems that people have in getting around
- It is not enough to be as safe as today: What is safe enough?

FOTs, deployments, use cases

- No common understanding of the terminology!
- FOTs
  - Data on technology + user behavior
- Deployments
  - Model deployments
    - Platform validation
  - Initial deployments
    - User benefits
- Use cases
  - Embedded operation of technological platforms and business models
  - Specific, advantageous locations with known policy environments

The clarity challenge

Three levels of definition

- Goals of the system (e.g. enhancing driving comfort, reducing travel time, improving user safety or broader traffic safety)
- Roles of the driver and the vehicle (SAE levels 0-5 deal with this aspect)
- Complexity of the operating environment

Differing philosophies

- Everything somewhere
- Something everywhere

- How safe is safe enough?
- Incrementalism

- Responsible capitalism
- Certification and regulation

Role(s) of research community

- Research support in key topic areas
  - Technology readiness
  - Human factors
  - "Data to understand the impacts of automated cars"
  - Cybersecurity
  - Legal and liability
  - Insurance
  - Public/private business models
  - User acceptance
- Purveyors of clarity
  - Understanding the levels of automation
  - Scrutiny of L3
- Conveners and deployers
  - Operating FOTs, deployments and use cases
  - Ecosystem cooperation
  - Public-private investment
  - Reducing uncertainty

Concluding remarks

- The collaboration is only beginning
  - Public-private
  - Academic role, cooperation between research groups
  - New constituencies (the full ecosystem)
  - EU-US
- 21st century mobility is voting with its feet
  - Private sector innovation
  - Consumer excitement
  - "Sooner rather than later"
  - New roles and opportunities for research community

Thank you!

Department for Transport

Connected & Automated Vehicles

Momentum in the UK

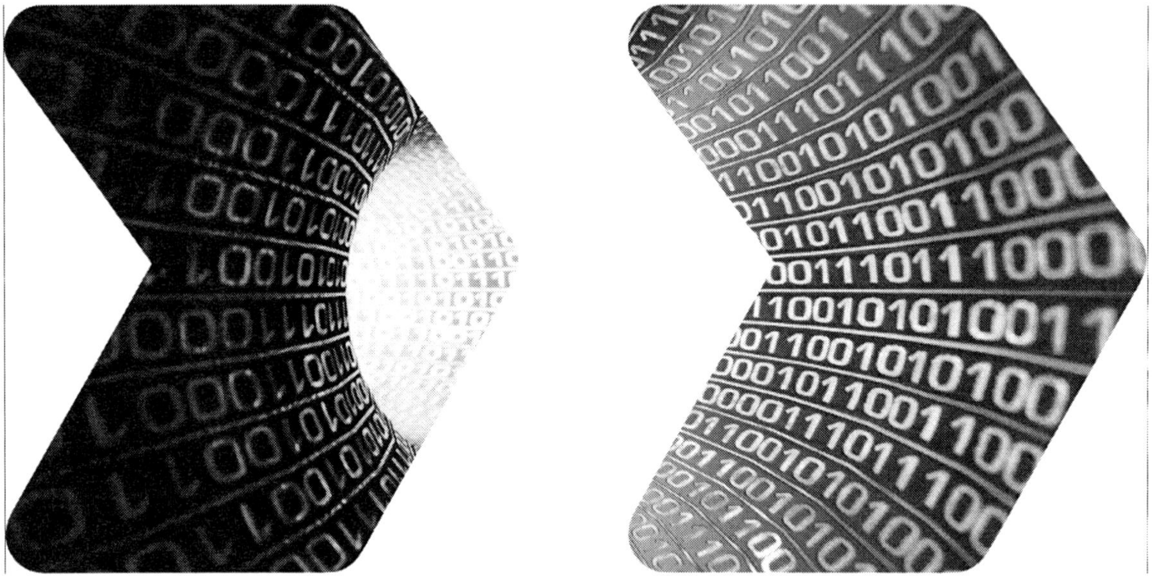

February 18

Department for Transport

There are times when we shape the future

February 18

Department for Transport

February 18

Department for Transport

All systems go

February 18

Department
for Transport

Department
for Transport

But to keep making progress…

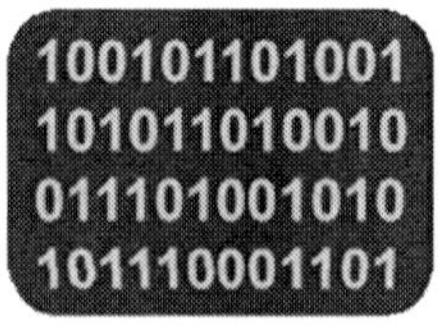

Department
for Transport

And finally…

2064

?

February 18

Department
for Transport

Thank you

@michael_hurwitz

February 18

CityMobil2

CITIES DEMONSTRATING AUTOMATED ROAD PASSENGER TRANSPORT

Automated road transport systems in European cities

Adriano Alessandrini
CTL – University of Rome

16/02/2015

www.citymobil2.eu

What is CityMobil2 and where we are

- Demonstrates ARTS
- Proposes a safety assessment procedure which can be applied in national legal frameworks for certification
- Foresees long term socio-economic effects of different automation scenarios

CityMobil2

Selected demo sites

CityMobil2

The news video of the Oristano demo

Video Presentation in
Original e-media

CityMobil2

December 2014 inauguration of La Rochelle CityMobil2 ARTS demonstration

CityMobil2 – Lausanne demo

Vantaa

7/23/2015 7

Preparing for Trikala

7/23/2015 CityMobil2 8

How many passengers have we carried so far?

- Oristano – 2580
- La Rochelle – 14660
- Lausanne – 4850
- Vantaa – 1610 (first 10 of 20 days)

- Total – 23700

7/23/2015 CityMobil2 9

What have we learned so far?

7/23/2015 CityMobil2 10

But what is it an ARTS and how does it differ from a Google car

- Individually used cars
- Door-to-doo
- From anywhere to anywhere
- Using any infrastructure
- No supervision

- Ride-shared vehicles
- Multimodal
- On pre-selected routes
- Only on certified (adapted) infras
- Full supervision

How does it translate in technical terms

- Dealing with failures
  - Redundancy vs. fail safe
- Dealing with other users
  - Forecast user behavior vs. integrated safety assessment
- Dealing with external problems
  - Recognizing police and roadworks vs. communicating with the control room

Using Chris' example

Source: Chris Urmson presentation at TED

CityMobil2

What the Google car needs to do

- The Google car needs to continue to work even when something is malfunctioning
  - Redundant energy supply
  - Redundant actuators ...
- It needs to detect everything on its own
  - The roadworks ahead (or eventually policemen and school busses)
- It predicts other user behavior

CityMobil2

What would we have done on vehicles

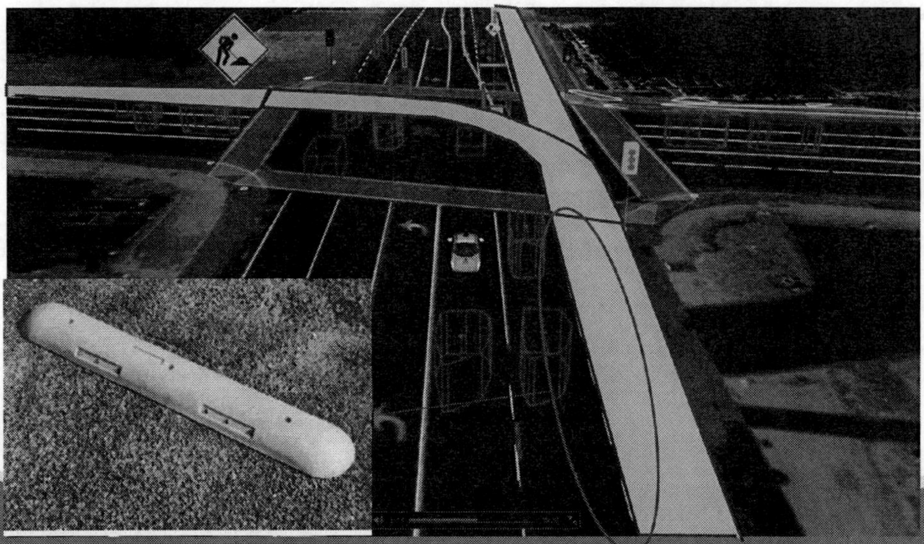

Elaboration on Chris' slide

CityMobil2

What would we have done to infrastructures

- Introduced an ARTS certified lane
  - with dedicated aspects in the light-cycle
  - with some protection
    - either hard barriers
    - or softer ones depending on the speed of the flow
- Allow manual vehicles on the ARTS lane forcing on them some rules
  - No-stopping, no-parking, no-overtaking
- Verified the distance to other moving and fixed obstacles and measured the time to collision and set the maximum speed of the vehicle on this basis
- Eventually put speed-bumps to slow down incoming vehicles on conflicting paths or sensors in the infrastructures to detect wrongdoing and preventing accidents

CityMobil2

What would we have done on vehicle and control

Elaboration on Chris' slide

What would we have done on vehicle and control

- Two dumb safety areas
  - The first (red) to trigger an emergency stop
  - The second (green) to induce a deceleration
- No software to forecast
- Pre-actuated brakes which need electricity to be de-actuated
- Respect of a pre-assigned trajectory
- Communication from the control room of any change of plan

How does it translate in use terms (1/2): automating individual mobility

With 70% of the kilometers full and 30% empty the occupancy rate would drop from 1.2 pax/car to 0.84 pax/car with 50% increase vehicle mileage

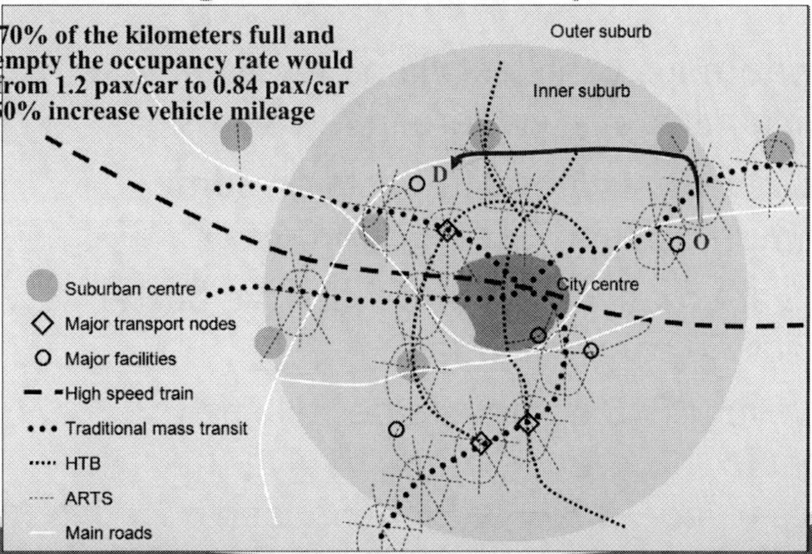

CityMobil2

How does it translate in use terms (2/2) automating collective mobility

Denser the demand (and concentrating either origin or destination) easier to ride-share with larger vehicles

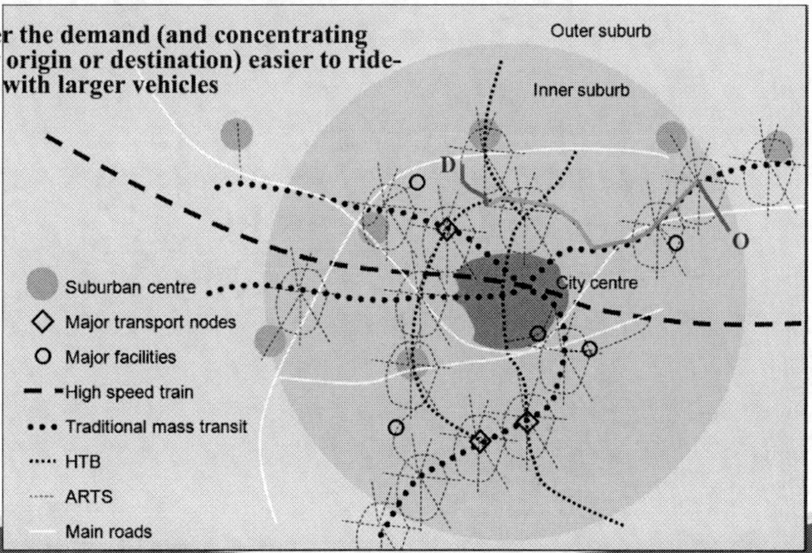

CityMobil2

Conclusions

- Autonomous vehicles will be finally able to drive almost everywhere in few years
- This could have good impacts on some communities and bad ones on others
- ARTS are ready today and will have better impacts in some realities
- However ARTS do need cities willing to invest in this dream; ready to give the right space to ride safely at higher speed

7/23/2015 CityMobil2 22

CityMobil2

CITIES DEMONSTRATING AUTOMATED ROAD PASSENGER TRANSPORT

Thank you for your attention

citymobil2@ctl.uniroma1.it

SEVENTH FRAMEWORK PROGRAMME

www.citymobil2.eu

Autonomous Vehicles in California

Stephanie Dougherty
Chief, Enterprise Planning & Performance
California Department of Motor Vehicles

July 22, 2015

California Legislation - Senate Bill 1298

- As soon as practicable, but no later than January 1, 2015, DMV must adopt regulations setting forth requirements for:
  - Manufacturers' testing of autonomous vehicles on public roadways
    - Regulations approved and adopted in May 2014
    - Effective on September 16, 2014
  - Operation of autonomous vehicles on public roadways
    - Under development

California's Definition of an Autonomous Vehicle

- Vehicle equipped with autonomous technology that has the capability to drive a vehicle without the active physical control or monitoring by a human operator
- Does not include vehicles equipped with one or more collision avoidance systems, such as electronic blind spot assistance, automated emergency braking systems, park assist, adaptive cruise control, lane keep assist, lane departure warning, and traffic jam and queuing assist

Testing on California Public Roadways

- $5 million in insurance, bond, or self-insurance
- Manufacturer has tested vehicle under controlled conditions that simulate, as closely as practicable, real word conditions and reasonably determined it is safe to operate the vehicle on public roads
- Test driver requirements:
  - No DUI, not an at-fault driver, and no more than 1 point
  - Successful completion of test driver training program
  - Employee, contractor, or designee of manufacturer
- Test driver must be seated in driver seat during testing
- Report any accident within 10 days and unanticipated disengagements of autonomous technology annually
- Testing permit valid for one year
- Vehicles excluded from testing: Commercial vehicles > 10,000 lbs GVW and motorcycles

Approved Testing Permits

Audi

Mercedes-Benz

Google

DELPHI

TESLA MOTORS

BOSCH
Invented for life

NISSAN

CRUISE

Operation on California Public Roadways

Development of Regulatory Package

- Statewide Steering Committee
- Consulting with University of California PATH
- Pre-notice workshops conducted in March 2014 and January 2015
- RFI to solicit input from third-party entities with an interest and capability in conducting functional safety reviews of autonomous vehicles

Regulatory Topics

- Certifications required by statute
- Conditions for operation
- Certification approaches

Permit to Deploy – The License to Drive

- Maintain $5 million in bond, insurance, or self-insurance
- Certify the autonomous technology meets and does not make inoperative any FMVSS
- Certify the autonomous vehicle has:
  - Mechanism to easily engage/disengage autonomous technology
  - Visual indicator that autonomous technology is engaged
  - System to alert the operator when a failure of the autonomous technology is detected
  - Separate mechanism to record sensor data 30 seconds prior to collision

Conditions for Operation

- Identify all areas where the vehicle can operate autonomously
- Identify commonly occurring restrictions on operation (snow, fog, rain, construction zones)
- Identify if vehicle is capable of operating without a driver inside
- Certify that the AV technology will obey traffic laws
- Identify what the vehicle will do if the autonomous technology fails

Vehicle Certification and Validation

- What should be certified?
  - Functional safety - Processes to identify hazards associated with the AV technology and procedures to mitigate those hazards
  - Behavioral competency - Ability to perform a minimal set of driving-related tasks and operate in the road traffic conditions the vehicle will regularly encounter
- By whom?
  - Federal government, state, manufacturer, third party
- As measured how?
- And verified how?

Thank You

DRIVE SWEDEN

Why?

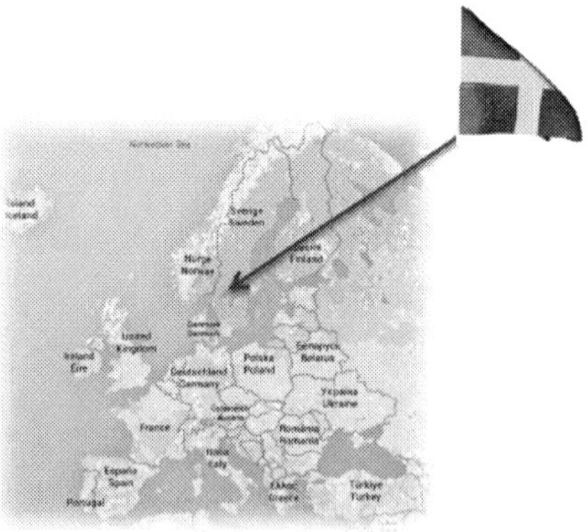

- Sweden in a nutshell:
  - 9.5 million people
  - Three leading vehicle OEMs
  - Solid track record in traffic safety
  - One major telecom player
  - Center for winter car testing
  - Early adopter mentality
  - Strong sustainability agenda
  - Successful startups (Skype, Spotify etc.)
  - Tradition of very good cooperation between government, industry and academia
  - Lots of automation activities already happening

DRIVE SWEDEN

How?

- The Swedish government has initiated several Strategic Innovation Programs (SIP), jointly financed by three government agencies plus a 50% industry contribution
  - The purpose of a SIP is to
    - create conditions for international competitiveness, and
    - to find sustainable solutions to global challenges for societies
  - Expected life cycle 12 years
  - Each SIP has the government's mandate to coordinate activities within its respective field

VINNOVA Swedish Energy Agency formas DRIVE SWEDEN

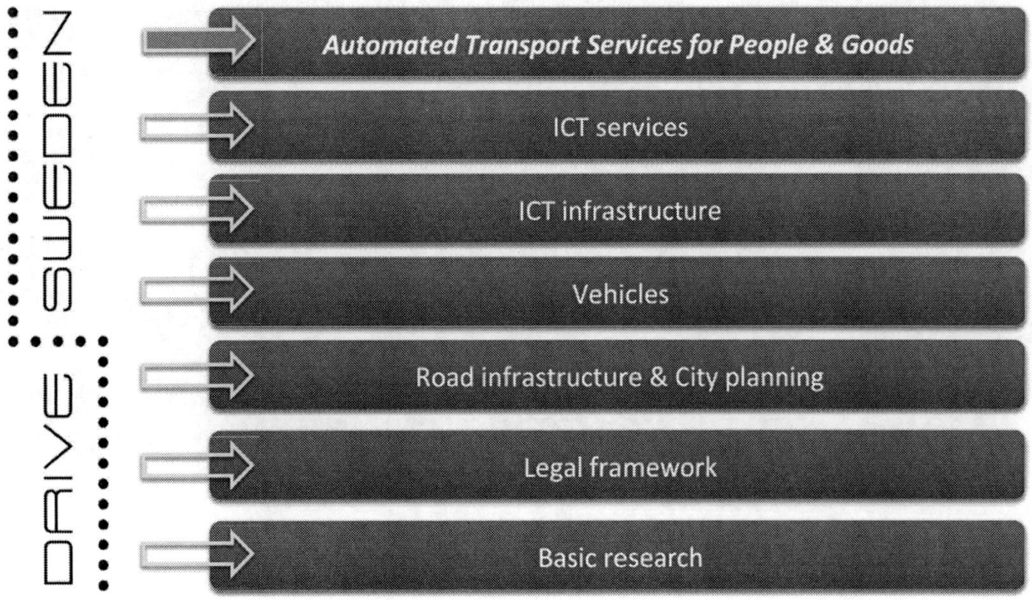

Who ?

- The founding members represent all relevant sectors:
  - Government & Cities
  - Vehicle manufacturers
  - ICT & Suppliers
  - Universities & Research institutes
  - Test facilities
- Drive Sweden is open for anyone to participate in

DRIVE SWEDEN

Some examples of projects

DRIVE SWEDEN

- World-first project on public roads
- 100 self-driving Volvo cars in real traffic
- Real customers behind the wheel
- $75 million investment
- Co-operation with Swedish authorities
- Project started 2013, real pilot in 2017

Drive Me

SELF-DRIVING CARS FOR SUSTAINABLE MOBILITY

City of Gothenburg | LINDHOLMEN SCIENCE PARK | VOLVO | TRANSPORT STYRELSEN | TRAFIKVERKET SWEDISH TRANSPORT ADMINISTRATION

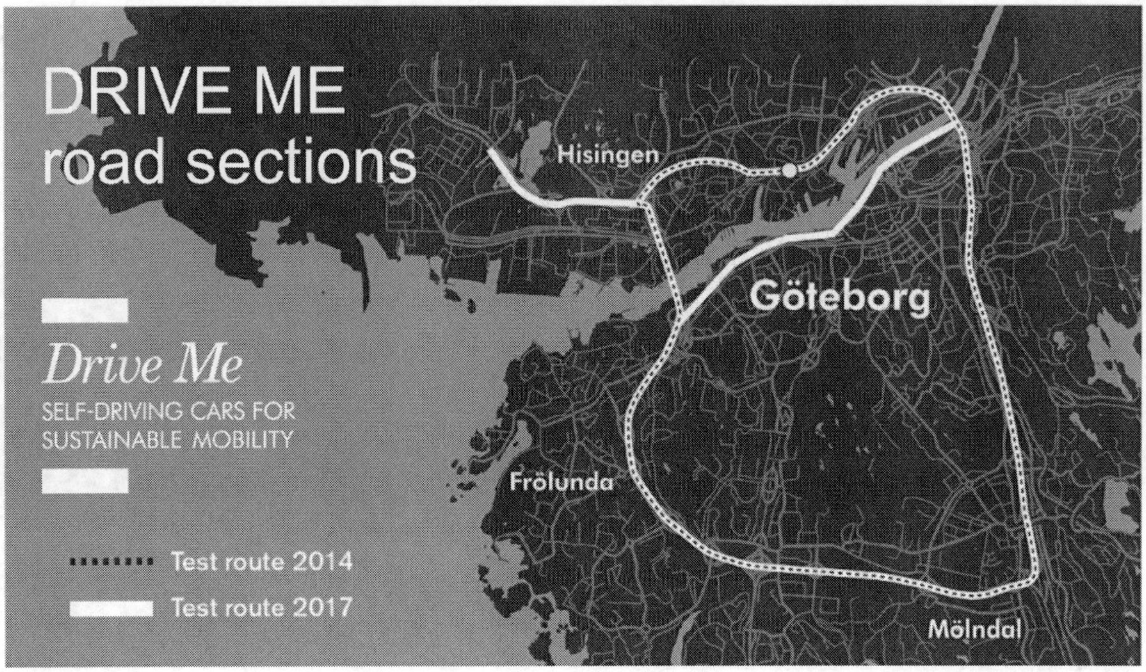

Focus areas

- Societal and economical benefits: improving traffic efficiency, traffic environment and road safety
- Infrastructure aspects
- Suitable traffic situations
- Customer expectations on autonomous vehicles
- How surrounding road users interact with a self-driving car
- Legal aspects

Cloud support for Automation

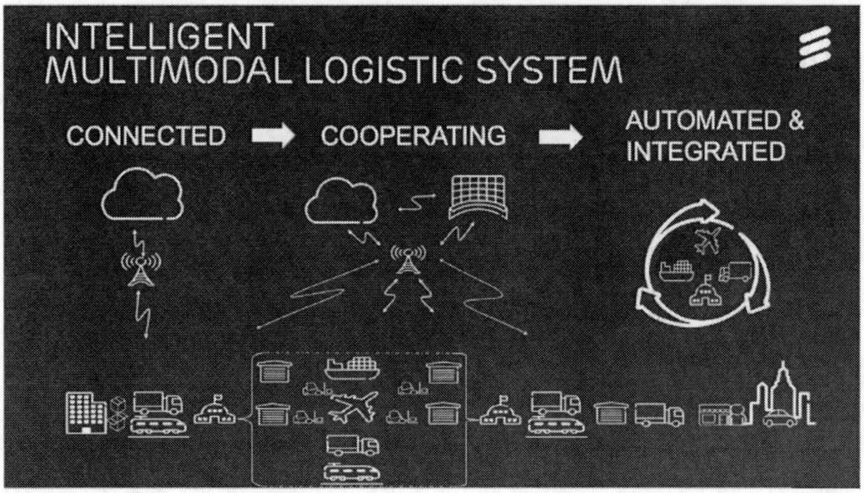

DRIVE SWEDEN

Truck platooning – a low hanging fruit

- SARTRE
- COMPANION
- FHWA BAA
- *FLOW4EU*

DRIVE SWEDEN

Next stop: Automation

DRIVE SWEDEN

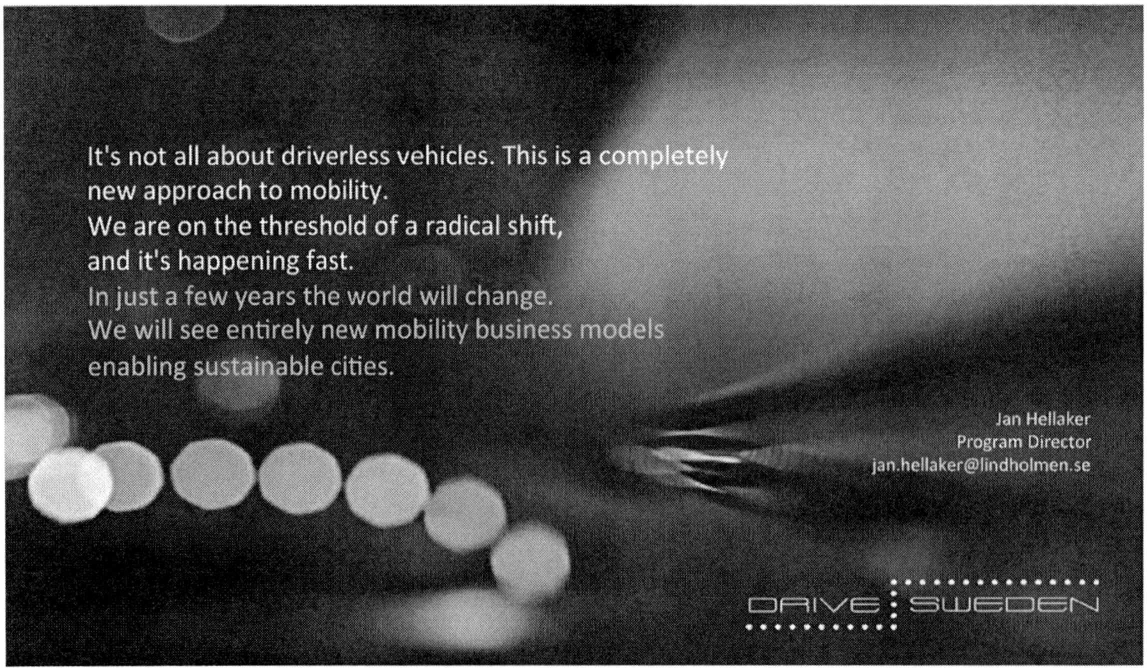

WORLD ECONOMIC FORUM

COMMITTED TO IMPROVING THE STATE OF THE WORLD

World Economic Forum

Multi-Stakeholder Cooperation on Self-Driving Vehicles

Alex Mitchell, Director, Head of Automotive Industry
22 July 2015

WORLD ECONOMIC FORUM

COMMITTED TO IMPROVING THE STATE OF THE WORLD

Why does the World Economic Forum care about self-driving vehicles?

Self-driving vehicles (SDVs) could bring tremendous societal benefits—effects in the US estimated at $1.3T (8% of GDP)

Annual benefits for fully autonomous vehicles in the US alone

Full automation	Unconditional, full-time performance of the driving task under all conditions that a human driver can manage
High automation	Part-time situation- or geography-dependent performance of the driving task of the automated driving system
Autonomy, self-driving	Common vernacular terms for both *full automation* and *high automation*
Connectivity	Here: Building block of incumbent approach to reap safety benefits along the way

30,000+ lives saved by avoided accidents

5.5B less hours spent in congestion

75B hours of regained commuting time

80% lane capacity improvement

40% fuel economy improvement

Source: SAE J3016 (for definitions), NHTSA, US Department of Transport, American Scientist, US Census Bureau, Morgan Stanley, Economist, World Economic Forum, BCG analysis

2

The World Economic Forum
The International Institution committed to improving the state of the world through public-private cooperation in the spirit of global citizenship
Non profit, international organization founded by Prof. Klaus Schwab in 1971.
Impartial: tied to no political, partisan or national interests.
Global: based in Geneva, with offices in New York, Beijing and Tokyo.
Belief: economic progress and social development are essential to creating a sustainable future.

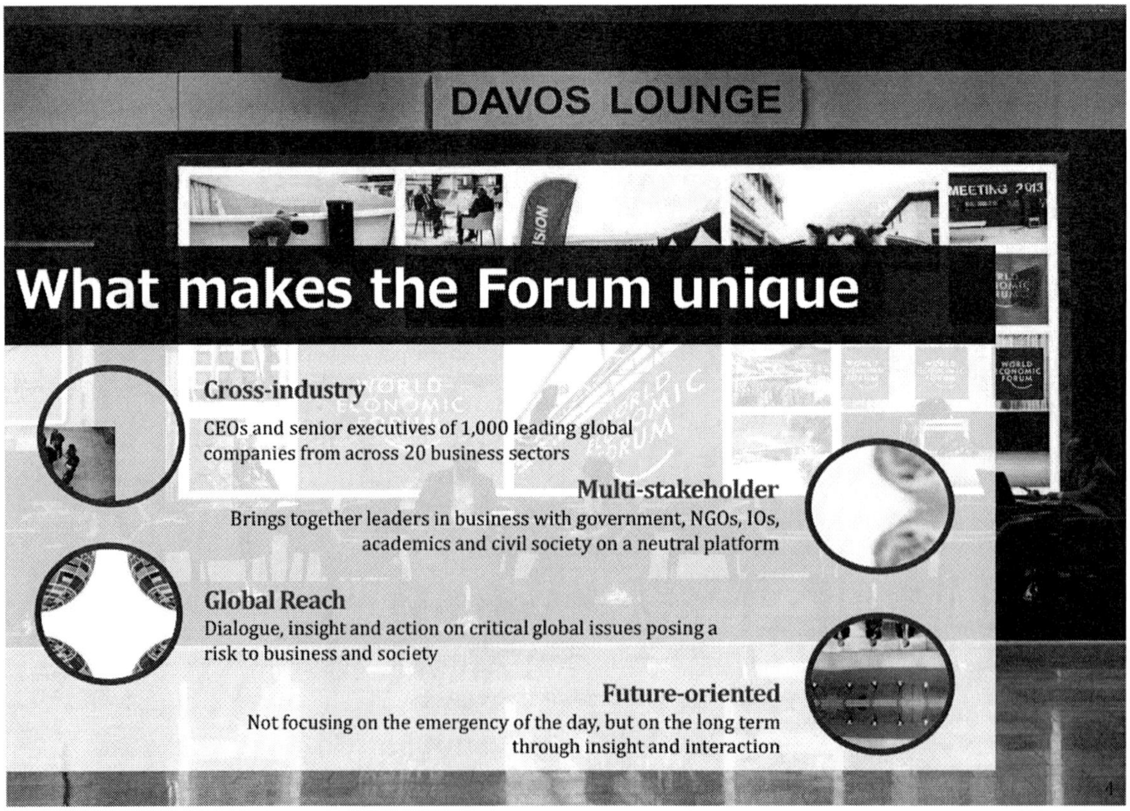
DAVOS LOUNGE
What makes the Forum unique
Cross-industry
CEOs and senior executives of 1,000 leading global companies from across 20 business sectors
Multi-stakeholder
Brings together leaders in business with government, NGOs, IOs, academics and civil society on a neutral platform
Global Reach
Dialogue, insight and action on critical global issues posing a risk to business and society
Future-oriented
Not focusing on the emergency of the day, but on the long term through insight and interaction
4

2014 SDV project work

WORLD ECONOMIC FORUM
COMMITTED TO IMPROVING THE STATE OF THE WORLD

Working group includes automotive, insurance, tech, and public sector

HYUNDAI
StateFarm
ZURICH
RENAULT NISSAN
DELPHI
Google
Audi
World Economic Forum jointly run initiative BCG
CISCO
GM
QUALCOMM
Continental
VOLVO
Government of the Netherlands

Objective: Draft a multi-stakeholder roadmap

Berlin	Detroit	New York

Evaluated most significant challenges	*Defined economic use cases*	*Developed a strategic roadmap*
May 20	Jul 9	Oct 20

Source: World Economic Forum; BCG analysis

5

Self-driving vehicles may be pursued in two different development approaches

WORLD ECONOMIC FORUM
COMMITTED TO IMPROVING THE STATE OF THE WORLD

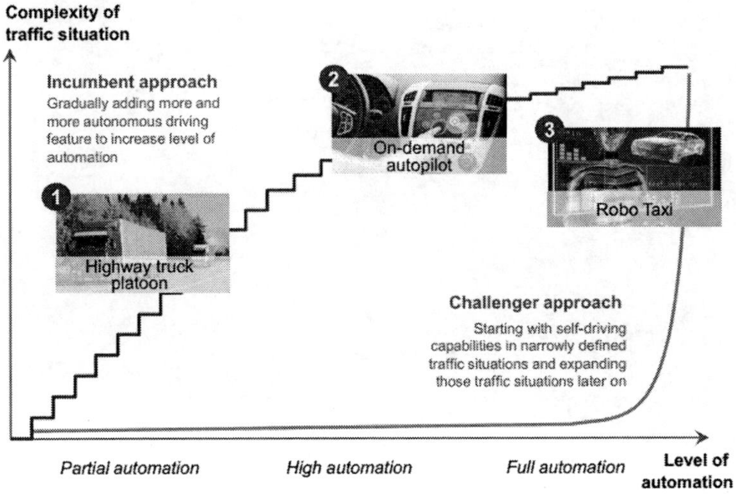

Comment

Incumbent approach
- Currently favored by incumbent OEMs
- Beliefs: Full automation takes time to materialize (societal attitude, technology maturity)
- Vehicle-to-vehicle comm. crucial building block
- Traditional introduction path: mass market launch starting with premium segment
- Focus on safety

Challenger approach
- Currently favored by tech players
- Beliefs: introduce tech early, improve on the go
- V2V communication optional
- Introduction possibly via fleet operators (e.g., closed campuses, cities, taxis)
- In addition to safety, focus on boosting convenience

Source: World Economic Forum; BCG analysis

6

The group selected and analyzed four significant challenges that require multi-stakeholder collaboration

Policy maker support	Liability shift	Making a safety case	Cyber-security

Support from policy makers, who are influenced by the general public and affected players, is crucial for SDV introduction.	*Liability obligations will likely shift if both human drivers and autopilots drive. The role of insurers may change*	*Existing (safety) standards and testing procedures do not account for SDV and their complex algorithms*	*SDV are highly connected systems that are prone to malicious hacking, potentially harm humans*
How can we ensure policy makers' support?	**How can we manage the liability shift for all affected stakeholders?**	**How can we develop new safety standards? How can we demonstrate safety?**	**How do we limit cybersecurity risks to an acceptable level?**

Source: World Economic Forum; BCG analysis

7

Regardless of the scenario, various players will have to get used to new roles in the eco-system

Regulator

- Gain skill regulating software, as opposed to just hardware
- Clarify the safety hurdle
- Be open to a liability shift away from the consumer

OEM/ T1Supplier

- Prepare for tech players being active in the space
- Prepare for liability residing with someone other than the human driver

Customer/ Consumer

- Cities: Prepare for sharing and autonomous to create "city as customer" for mobility in a city

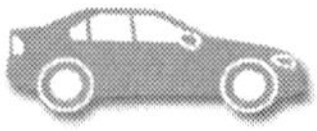

Self-driving vehicle

Tech Firm

- Adapt to longer product cycles and high safety regulations
- Bring cyber-security best practices from other industries (e.g., banking)

Insurer

- Accept potentially lower auto premiums over time
- Prepare for liability residing with someone other than the human driver

Source: World Economic Forum; BCG analysis

8

WORLD ECONOMIC FORUM
COMMITTED TO IMPROVING THE STATE OF THE WORLD

2015 Project Work: Focus on the urban application

This initiative allows cities and industry to assess what self-driving vehicles mean for urban spaces in a real life trial, confined to an area of the city.

This includes a focus on shared, on-demand, driverless transport.

The objective is to provide a blueprint for cities to make new mobility options available to citizens

9

WORLD ECONOMIC FORUM
COMMITTED TO IMPROVING THE STATE OF THE WORLD

2015 Project Work: 3 major workstreams

1 Understand future customers	2 Develop future urban mobility models	3 Facilitate implementation

Consumers: Conduct survey of 5-10k consumers globally to test new mobility scenarios

Cities: Interviews w/ policy makers in cities; Understand key challenges

Develop and calculate 2–3 operating and business models
— Model economics for consumers, mobility players and cities

Involve 2–3 *core cities* early on to devise road-map for most 2–3 promising scenarios

Source: World Economic Forum; BCG analysis

10

Automated Vehicles and Public Perception

J.D. POWER
McGRAW HILL FINANCIAL

AUTOMATED VEHICLES SYMPOSIUM 2015

DRIVERS. VEHICLES. INFRASTRUCTURE.

July 22, 2015

Kristin Kolodge | kristin.kolodge@jdpa.com | 248.680.6446

I want a faster horse...

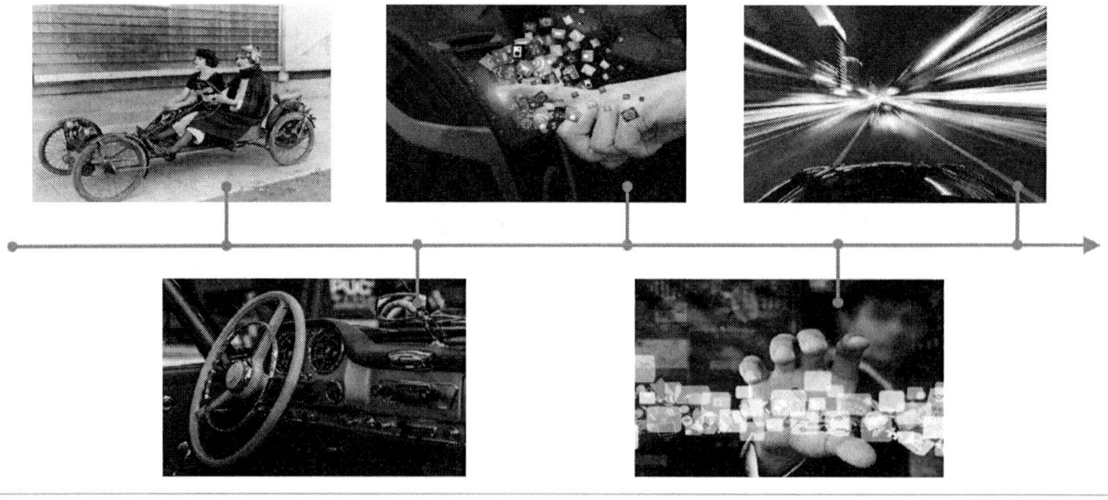

J.D. POWER
McGRAW HILL FINANCIAL

Vehicles with Blind Spot Monitoring scored higher in APEAL than those that did not have the feature

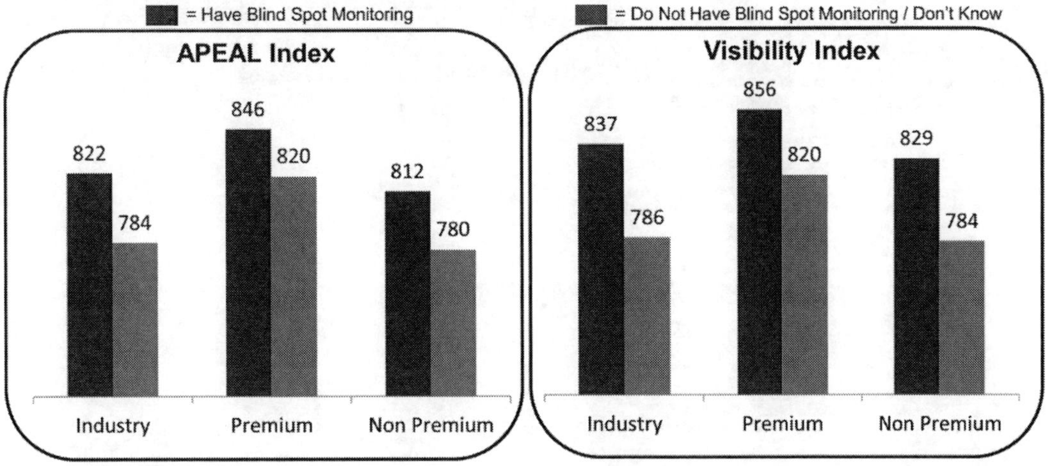

Source: J.D. Power 2015 U.S. APEAL Study

Vision assistance systems rate high for desirability and usage in vehicles today and are preferred on consumer's next purchase

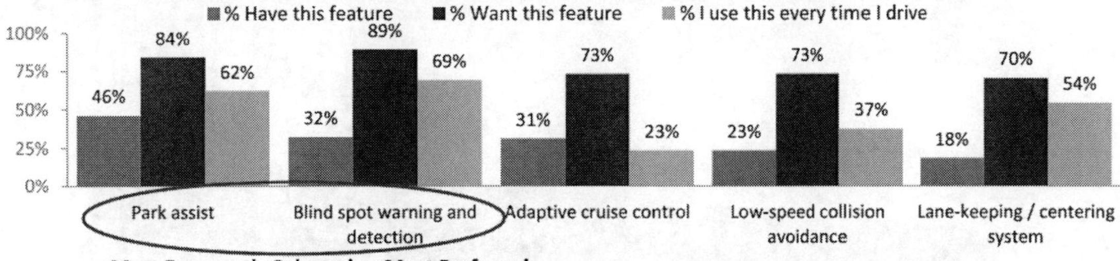

Most Frequently Selected as Most Preferred

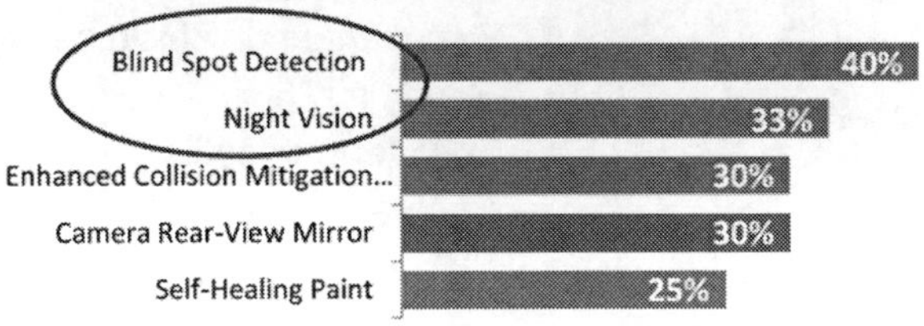

Respondents are consistently demonstrating their desire for assistance while driving as supported by J.D. Power's DrIVE 1st Gear and also Tech Choice, in which Blind Spot Detection and Prevention was the most preferred technology , even at a price of $750

Source: J.D. Power 2015 U.S. DrIVE 1st Gear Study and J.D. Power 2015 U.S. Tech Choice Study

Collision Protection Preference (MaxDiff Score)

Auto braking & steering common theme

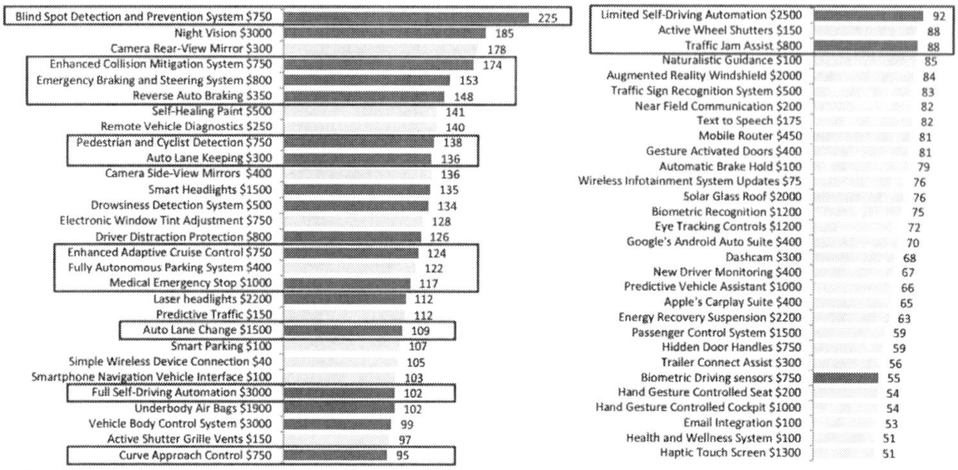

Source: J.D. Power 2015 U.S. Tech Choice Study

J.D. POWER McGRAW HILL FINANCIAL

Consumer Shift Toward Collision Protection Technologies

- Occupant and vehicle safety
- Ready for "help"
- Long term benefits
- Enabler

- In-vehicle experience
- Driver distraction
- Value add benefit
- Cost (in a world of "free")

Source: J.D. Power 2015 U.S. Tech Choice Study

J.D. POWER McGRAW HILL FINANCIAL

Who wants a Collision Protection Avoidance Technology on their next vehicle?

Interest by Age

Generation Y			Generation X
87% (2025)	$89k	79%	$108k

Generation Y: 87% – 2025 – $89k

Generation X: 79% – $108k

Boomers: 76% – $123k

Pre-Boomers: 81% – $97k

Source: J.D. Power 2015 U.S. Tech Choice Study

J.D. POWER
McGRAW HILL FINANCIAL

Building Blocks for Self-Driving Automation

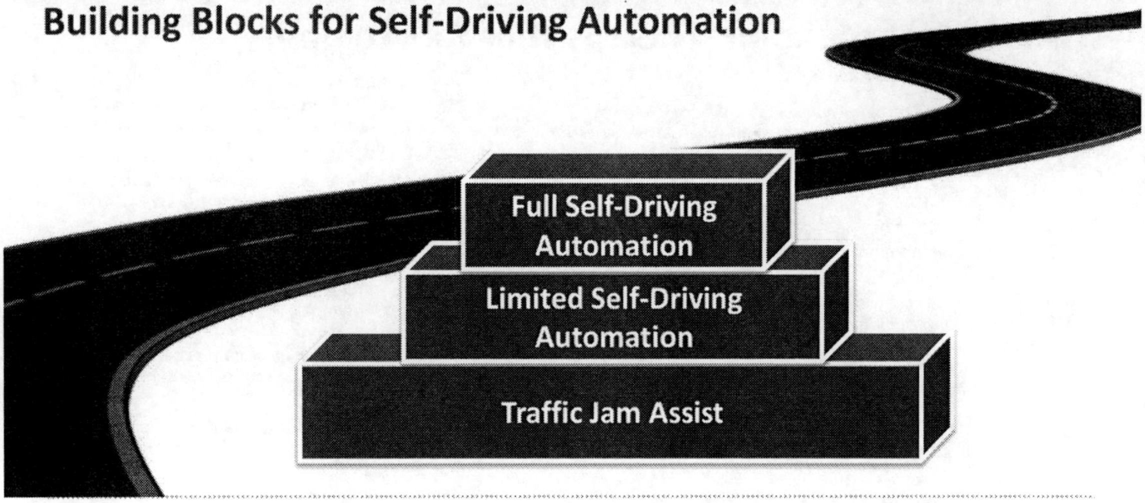

As the industry builds toward the future where Self-Driving vehicles are a reality, it is important to understand how different technologies will influence that movement and what consumers think of those technologies as we move forward toward acceptance.

J.D. POWER
McGRAW HILL FINANCIAL

On the Road to Self-Driving Vehicles

Generations differ in their preference for these technologies

Preference (MaxDiff Score) by Generation

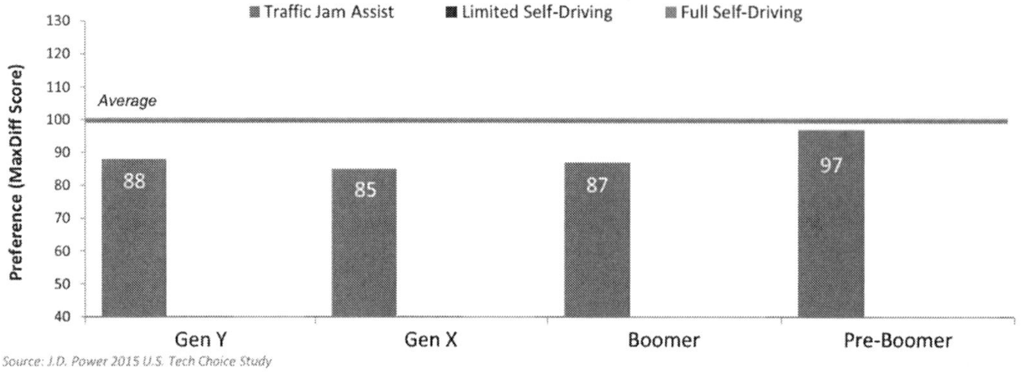

Source: J.D. Power 2015 U.S. Tech Choice Study

Gen Y and Gen X seem to prefer the more complex and advanced Self-Driving technologies that can handle all or some of the driving, while the older generations prefer to ease into full Self-Driving showing higher preference for Traffic Jam Assist than the younger generations.

J.D. POWER McGRAW HILL FINANCIAL

Two sides of the coin—There is still work to do

"The potential of self driving cars is awesome."

"Instead of spending time fuming at the wheel, 'drivers' would be free to get on with work or watch a film."

"GIMME A SELF DRIVING CAR NOAW !!!!!! No more traffic jams, sleeping in the car !"

"Self driving cars are an awesome idea. Think about it; it can drive to where you are - so the car can do all the dull things like finding a parking space for you."

"Smart cars for drunk drivers! Let's take drunk drivers off the roads and safely home in smart cars."

"Google's self-driving car is already better than most drivers. If they can figure out winter driving, there's nothing stopping them."

*"Self-driving cars are pretty baller. One thing I've always wondered is what happens **if the software fails and your car hits something, who would be at fault?**"*

*"I think that a possible problem with self-driving cars is that **if the software crashes, so do you.**"*

*"A self driving car is a great idea. **A self driving car without manual override, not so much.**"*

*"**Driving is such a fun experience** why do people want self driving cars"*

*"**This is part of the future that I would resist, even though I see the benefits.** The cars are safer, more economical, more energy efficient. They also look absurd, take control away from the individual, limit freedom, and only resemble an "automobile" in the most basic functional sense."*

Source: J.D. Power 2015 U.S. Tech Choice Study Social Media Analysis

J.D. POWER McGRAW HILL FINANCIAL

Redefinition of Quality

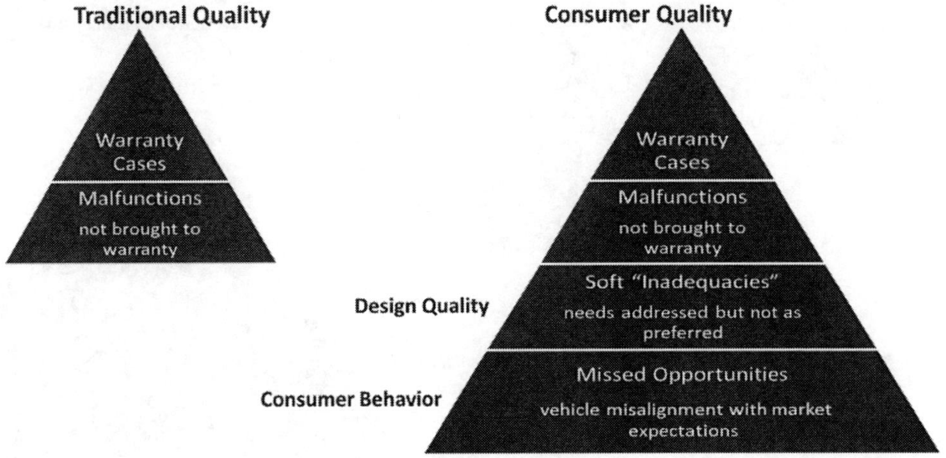

Source: J.D. Power 1987 through 2015 U.S. Initial Quality Study (IQS)

J.D. POWER McGRAW HILL FINANCIAL

While consumers are finding this technology helpful, only 39% reported that the dealer explained the feature to them. Further, false positives and false negatives are risk factors that can erode trust in the system

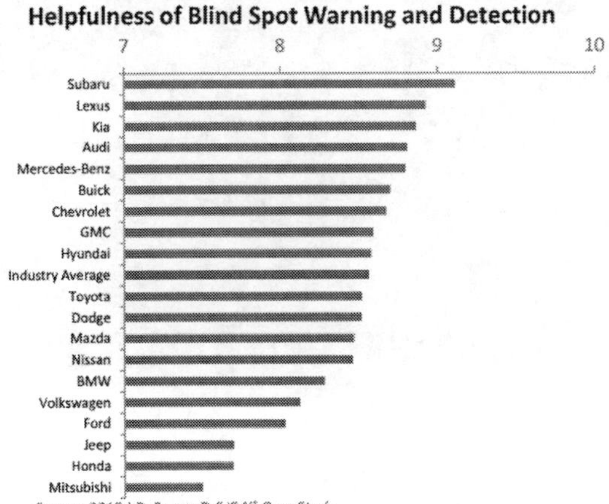

Source: 2015 J.D. Power DrIVE 1st Gear Study

"Sometimes the system fails to detect a car in my blind spot. There are also a lot of false positives when I'm driving next to a barrier or in a double turn lane (but the double lane one is unavoidable)."

"Alerts when nothing is nearby."

"Not even sure if I have it or not."

"Took me 6 weeks to even know how to use. The factory default was set to off!!! And the dealer didn't explain the feature to me."

J.D. POWER McGRAW HILL FINANCIAL

Evolution of Consumer Quality Pyramid

Trust Issues Arise

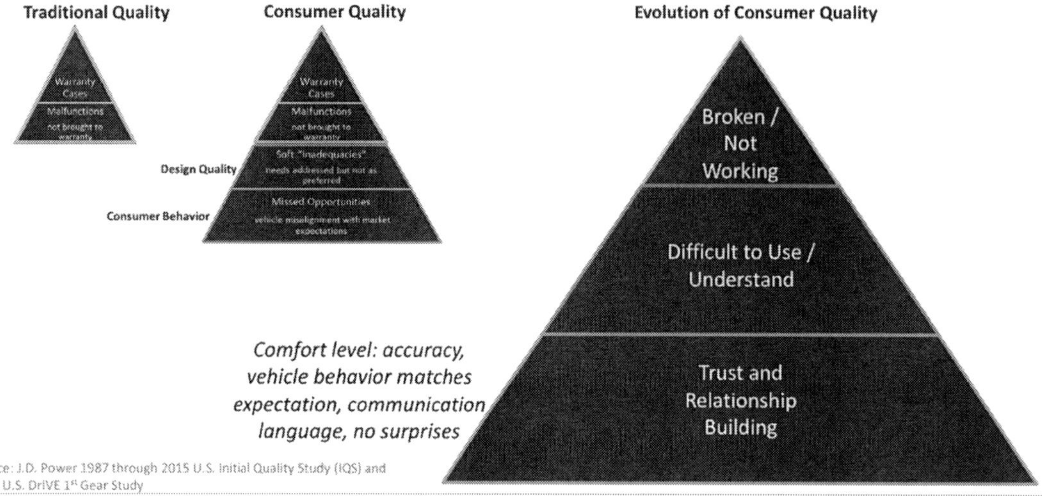

Source: J.D. Power 1987 through 2015 U.S. Initial Quality Study (IQS) and 2015 U.S. DrIVE 1st Gear Study

J.D. POWER
McGRAW HILL FINANCIAL

Thank You

J.D. POWER
McGRAW HILL FINANCIAL

J.D. POWER

McGRAW HILL FINANCIAL

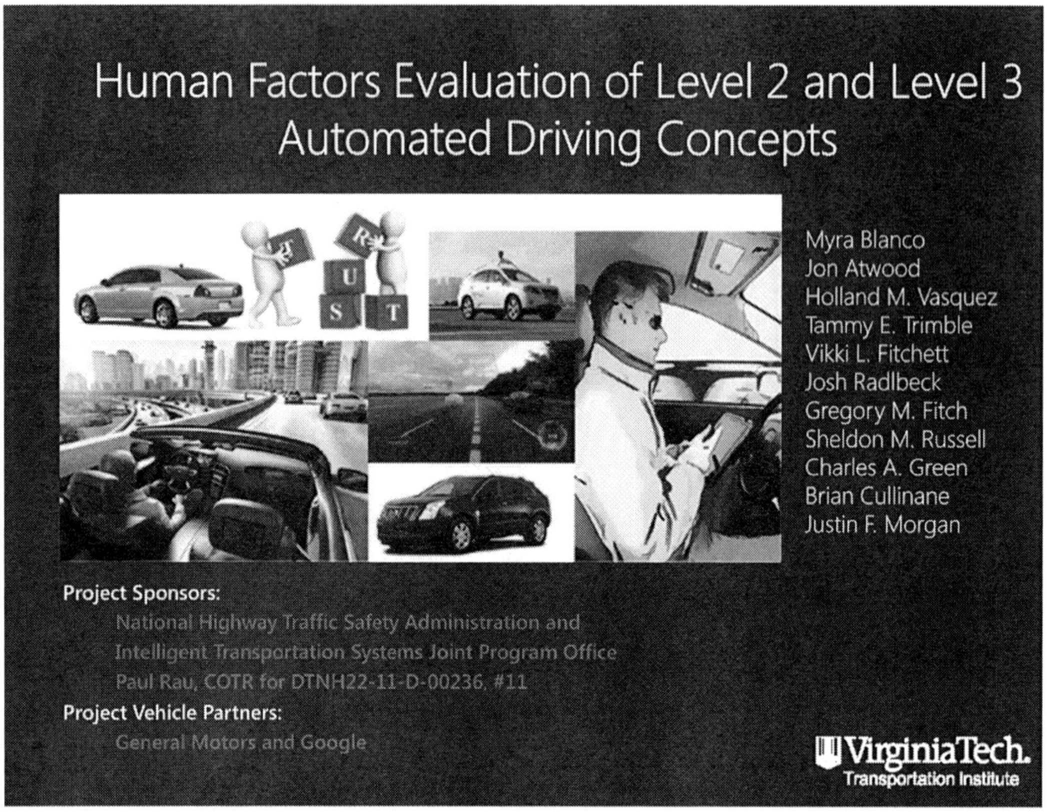
Human Factors Evaluation of Level 2 and Level 3 Automated Driving Concepts
Myra Blanco
Jon Atwood
Holland M. Vasquez
Tammy E. Trimble
Vikki L. Fitchett
Josh Radlbeck
Gregory M. Fitch
Sheldon M. Russell
Charles A. Green
Brian Cullinane
Justin F. Morgan
Project Sponsors:
National Highway Traffic Safety Administration and
Intelligent Transportation Systems Joint Program Office
Paul Rau, COTR for DTNH22-11-D-00236, #11
Project Vehicle Partners:
General Motors and Google
VirginiaTech.
Transportation Institute

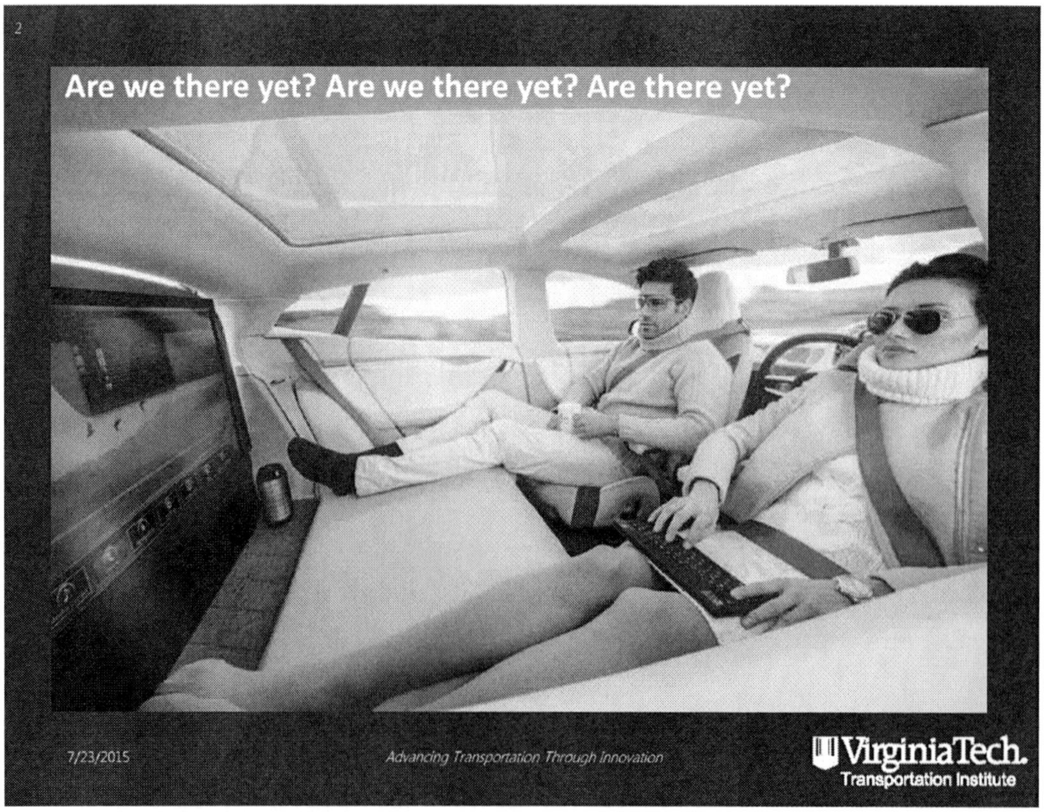
2
Are we there yet? Are we there yet? Are there yet?
7/23/2015
Advancing Transportation Through Innovation
VirginiaTech.
Transportation Institute

3

Overview of Three Experiments

Experiment 1 – L2

Alert Type (within subject)
- Cautionary
- Staged
- Imminent

Alert Modality (within)
- Unimodal
- Multimodal

25 participants

One 90-min session

Experiment 2 – L2

Driving Session (within)

Event Type (within)
- Alert
- No Alert
- No Lane Drift

Prompt Condition (between subjects)
- 2-s
- 7-s
- No prompt

56 participants

Three 60-min sessions

Experiment 3 – L3

Driving Session (within)

Alert Type (within)
- Staged
- Imminent – External Threat
- Imminent – No External Threat

25 participants

Three 30-min sessions

7/23/2015 Advancing Transportation Through Innovation VirginiaTech. Transportation Institute

4

Vehicles and Partners

Experiment 1 (L2 ADS)

2009 Chevy Malibu

GM

Experiment 2 (L2 ADS)

2010 Cadillac SRX

GM

Experiment 3 (L3 ADS)

2012 Lexus RX450h

Google

5

Dependent Variables

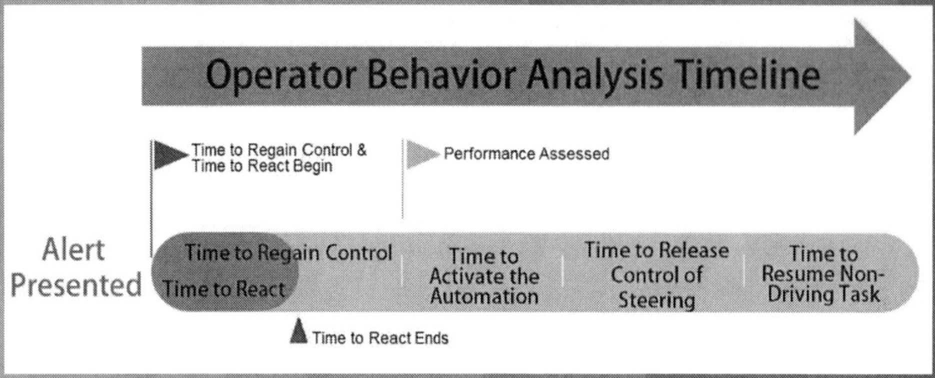

7/23/2015 Advancing Transportation Through Innovation

Virginia Tech. Transportation Institute

6

Experiment 2

- 56 participants; mean age = 41 yrs.
- Investigated L2 attention prompt effectiveness
- Drivers experienced 2-s, 7-s, or no prompts
  - Prompts progression
    - Stage 1: Visual
    - Stage 2: Visual + haptic
    - Stage 3: Visual + haptic + auditory

7/23/2015 Advancing Transportation Through Innovation

Virginia Tech. Transportation Institute

7

Experiment 2

- Three 1-hour driving sessions
- Asus Nexus 7 tablet computer was provided to participants
- In-vehicle experimenter gave a series of navigation, email, and web-browsing tasks
- 30 tasks in each category, potential of 90 tasks in all

7/23/2015 *Advancing Transportation Through Innovation* VirginiaTech. Transportation Institute

8

Driving-related Glance Time (Attention to Roadway)

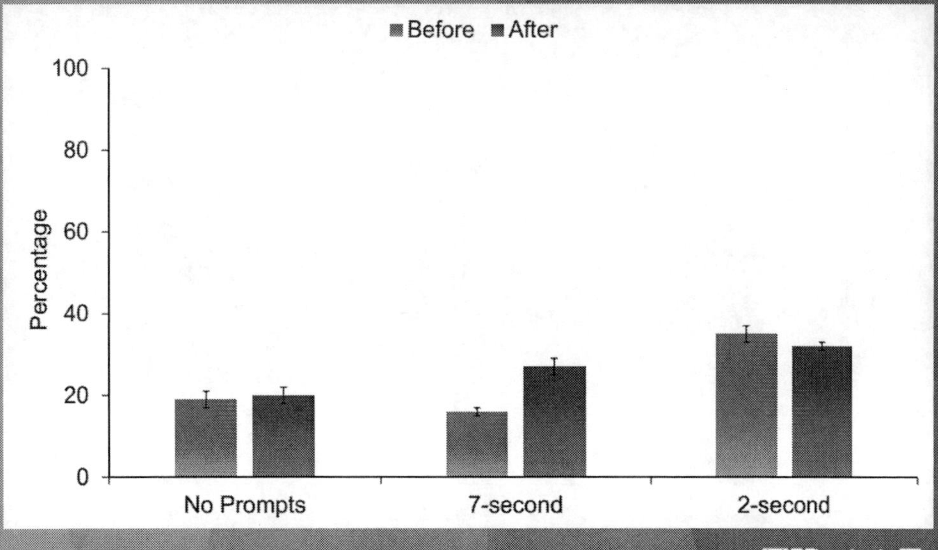

7/23/2015 *Advancing Transportation Through Innovation* VirginiaTech. Transportation Institute

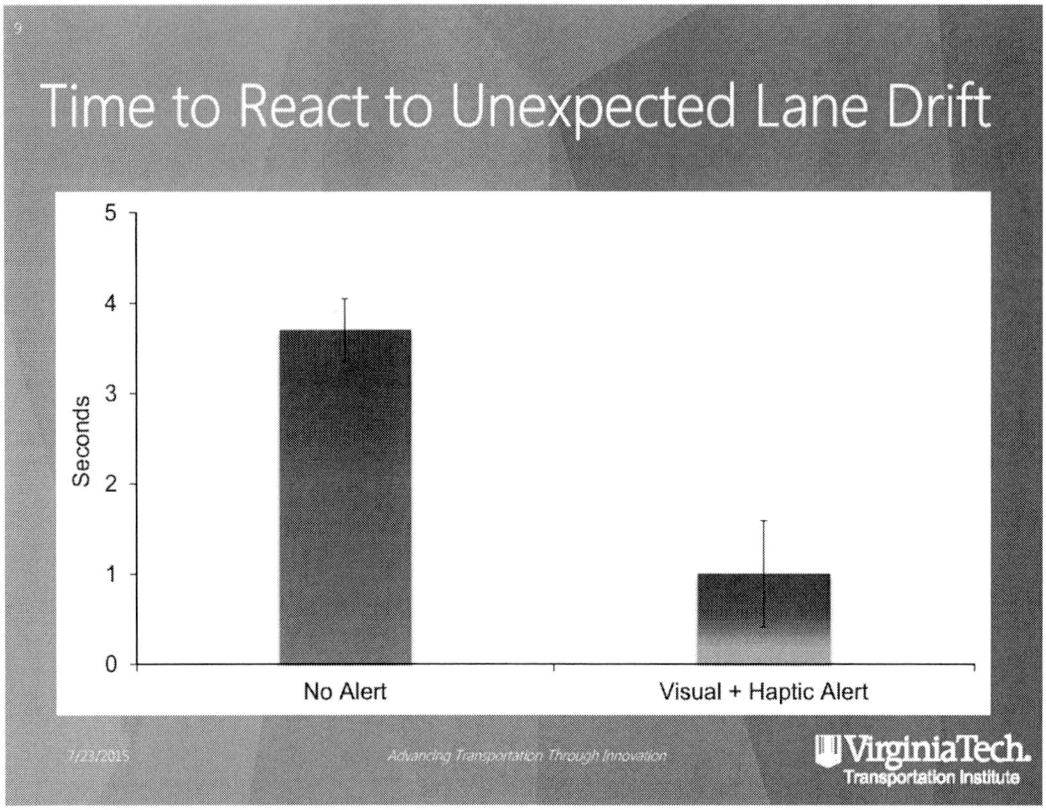
9
Time to React to Unexpected Lane Drift
Seconds
5
4
3
2
1
0
No Alert
Visual + Haptic Alert
7/23/2015
Advancing Transportation Through Innovation
VirginiaTech.
Transportation Institute

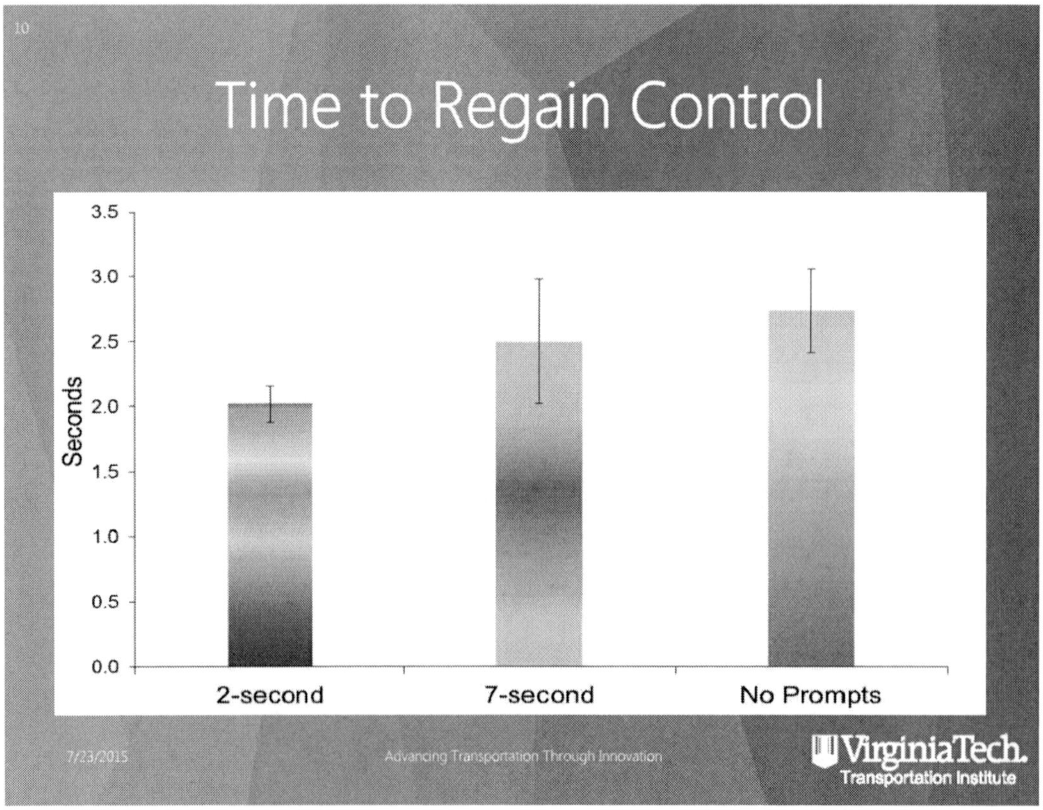
10
Time to Regain Control
Seconds
3.5
3.0
2.5
2.0
1.5
1.0
0.5
0.0
2-second
7-second
No Prompts
7/23/2015
Advancing Transportation Through Innovation
VirginiaTech.
Transportation Institute

11

Experiment 3

- 25 participants; mean age = 38.8 yrs.
- Investigated L3 Take-Over Request Effectiveness
- Drivers received one alert per 30-minute session
  - Staged
  - Imminent – No External Threat
  - Imminent – External Threat (i.e., box on road)

7/23/2015 *Advancing Transportation Through Innovation* VirginiaTech. Transportation Institute

12

Experiment 3

- Three 30-min driving sessions
- Participants were allowed to perform tasks and access Internet on Asus Nexus 7 tablet and use their personal smartphone as they wished
- Tablet was pre-loaded with movies, games
- Tasks to be done only when L3 automation was activated

7/23/2015 *Advancing Transportation Through Innovation* VirginiaTech. Transportation Institute

13

Time to Regain Control (Staged Alert)

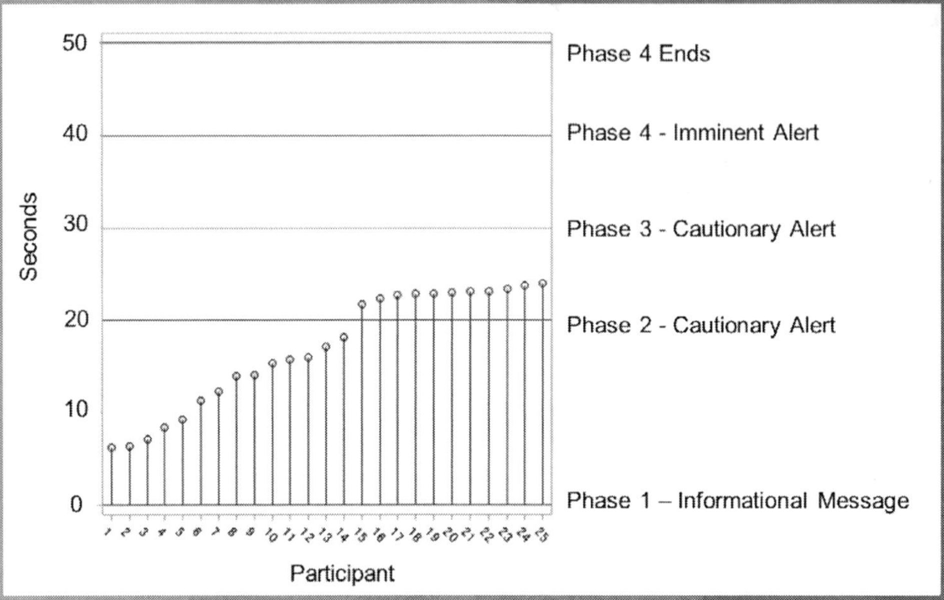

14

Key Takeaways

- Take Over Request
  - Most effective hand-off strategies were those that incorporated nonvisual components
    - Effective countermeasures to primary task reversals when drivers performed non-driving tasks
- Regain Control
  - L2 mean of 1.3 s (S.E. = 0.1 s)
    - Imminent visual and haptic alert
  - L3 mean of 2.3 s (S.E. = 0.2 s)
    - Imminent visual plus auditory alert
- Trust
  - High trust in automation for both levels of automation but calibrated
    - Trust was reduced after events where something occurred unannounced

7/23/2015 *Advancing Transportation Through Innovation* VirginiaTech. Transportation Institute

15

Vehicle Automation Theories

- Primary Task Reversal
- Alert Annoyance Habituation

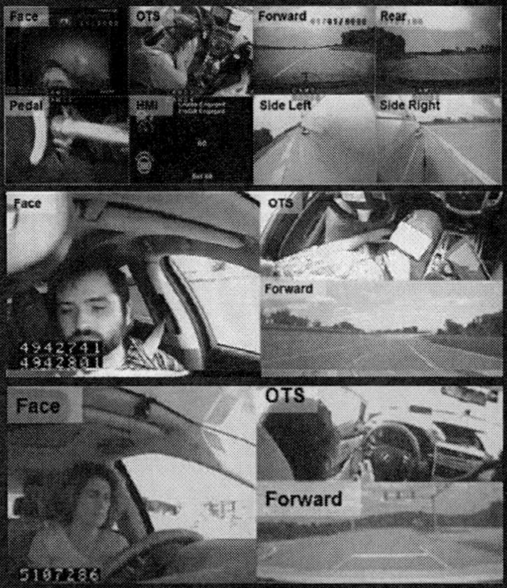

7/23/2015 *Advancing Transportation Through Innovation* VirginiaTech. Transportation Institute

16

Primary Task Reversal

- Full-priority shift from driving-related task to non-driving tasks
  - Non-driving tasks becomes primary task demoting controlling the vehicle to secondary task
  - Readiness to respond to driving-related prompts and alerts can be delayed because operators feel obliged to complete non-driving task first

7/23/2015 *Advancing Transportation Through Innovation* VirginiaTech. Transportation Institute

17

Alert Annoyance Habituation

- Operators can weigh non-driving task as more urgent if the TOR alert's urgency is low
- Operators can weigh the non-driving task as less urgent if the TOR alert urgency is high
- Need HMIs that balance conspicuity, urgency, and annoyance

7/23/2015 Advancing Transportation Through Innovation VirginiaTech. Transportation Institute

18

Acknowledgments

Many thanks to NHTSA and ITS JPO, the project's sponsors; to Dr. Paul Rau, Contracting Officer's Technical Representative; and to our partners and stakeholder committee members!

Blanco, M., Atwood, J., Vasquez, H.M., Trimble, T.E., Fitchett, V.L., Radlbeck, J., Fitch, G.M., Russell, S.M., Green, C.A., Cullinane, B., & Morgan, J.F. (In Press). *Human factors evaluation of level 2 and level 3 automated driving concepts: Final report.* Washington, DC: National Highway Traffic Safety Administration.

7/23/2015 *Advancing Transportation Through Innovation* VirginiaTech. Transportation Institute

19

Questions

VirginiaTech.
Transportation Institute

Myra Blanco
mblanco@vtti.vt.edu

From "Human Factors" to "User Experience" for level 3 car automation:

the "2-in-1" Mobius concept for intuitive driving

Patrice Reilhac
Innovation & Collaborative Research Director
Comfort & Driving Assistance Business Group
AVS 2015, Ann Harbor USA

Valeo

Valeo Mobius concept in the Context of Automation Levels (SAE)

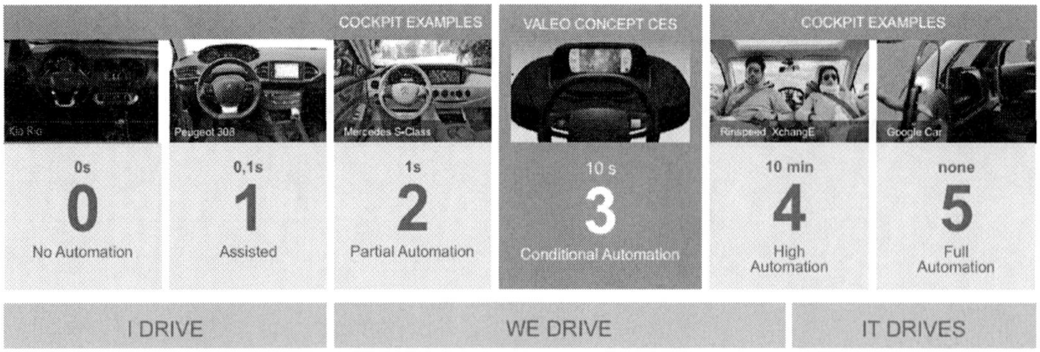

Valeo

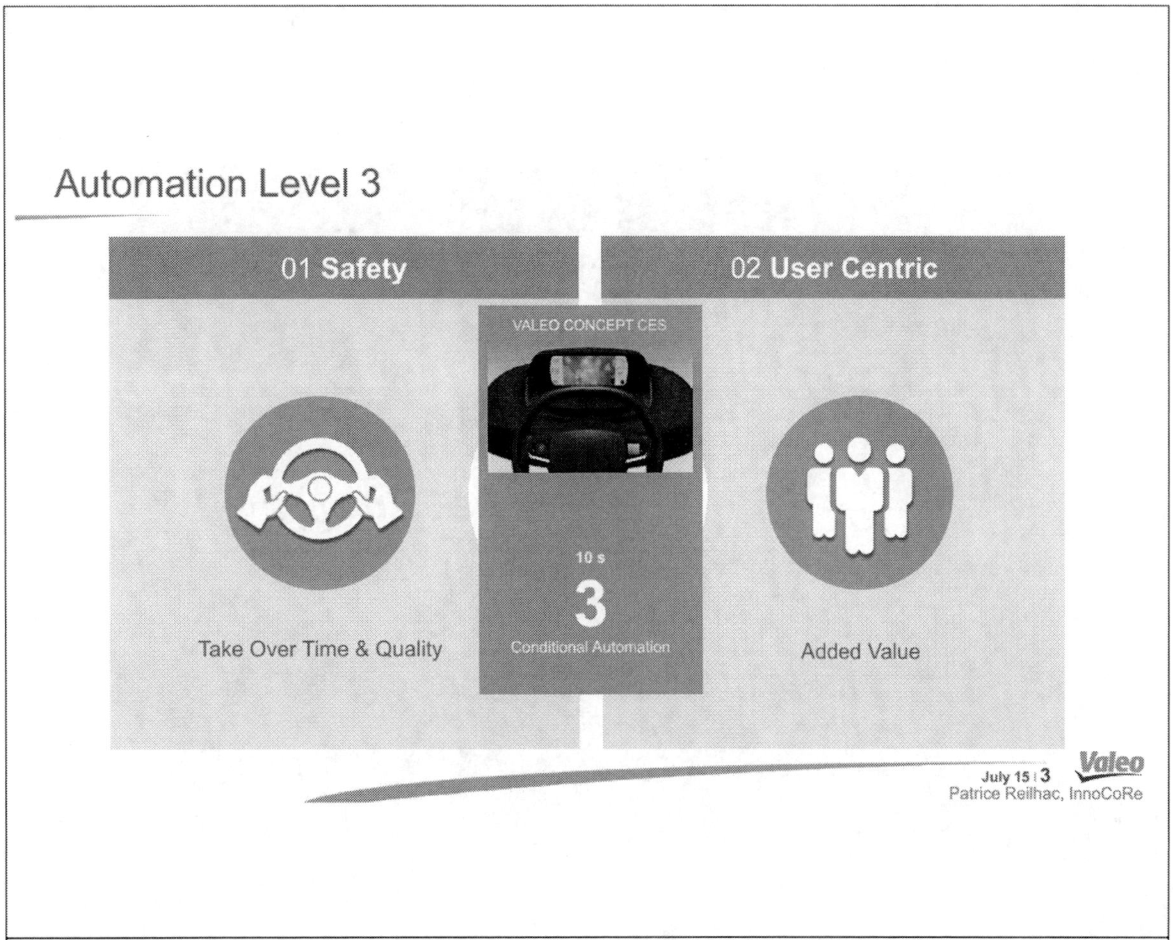
Automation Level 3
01 Safety
02 User Centric
VALEO CONCEPT CES
10 s
3
Conditional Automation
Take Over Time & Quality
Added Value
Valeo
July 15 | 3
Patrice Reilhac, InnoCoRe

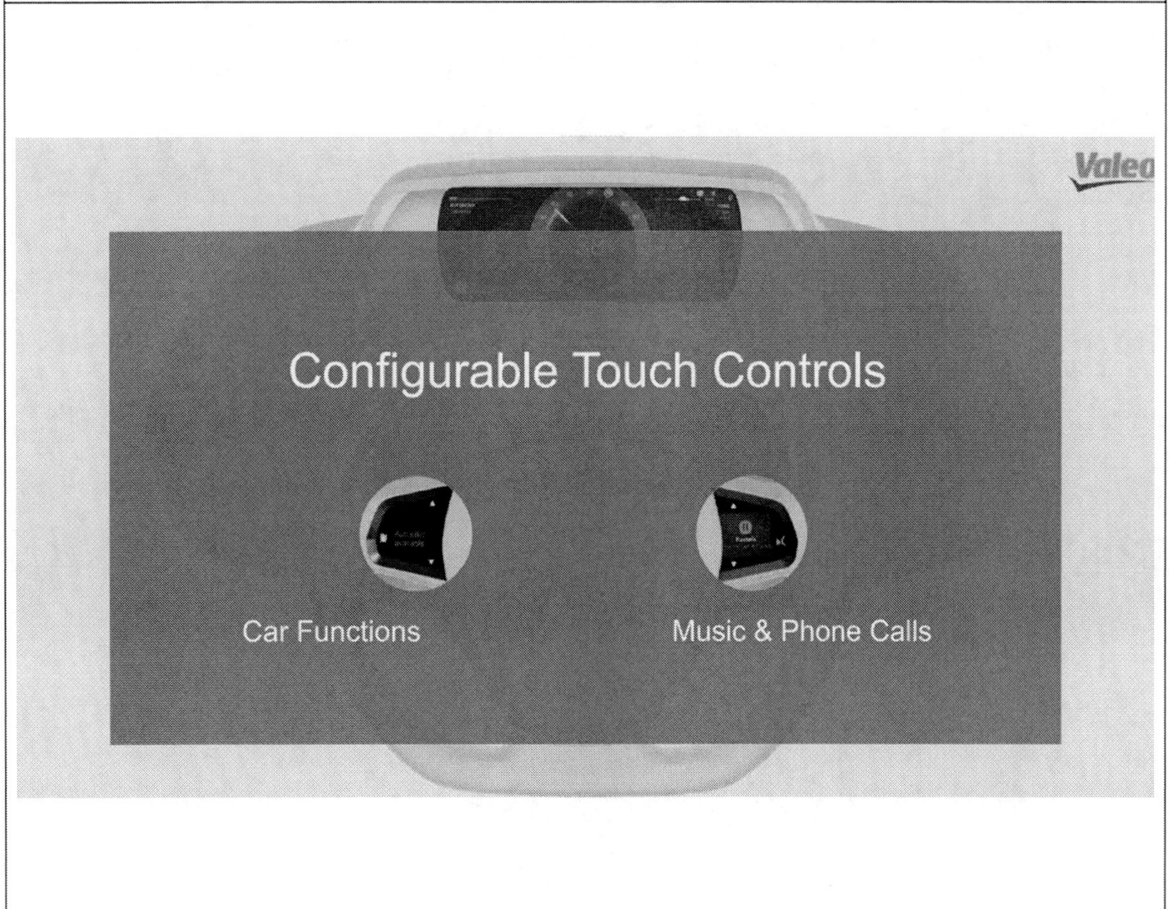
Valeo
Configurable Touch Controls
Car Functions
Music & Phone Calls

AUTOPILOT
Valeo
38
19.2
User Experience:
Mirroring of Smart Device
in the Instrument Cluster
via HDMI or WiFi

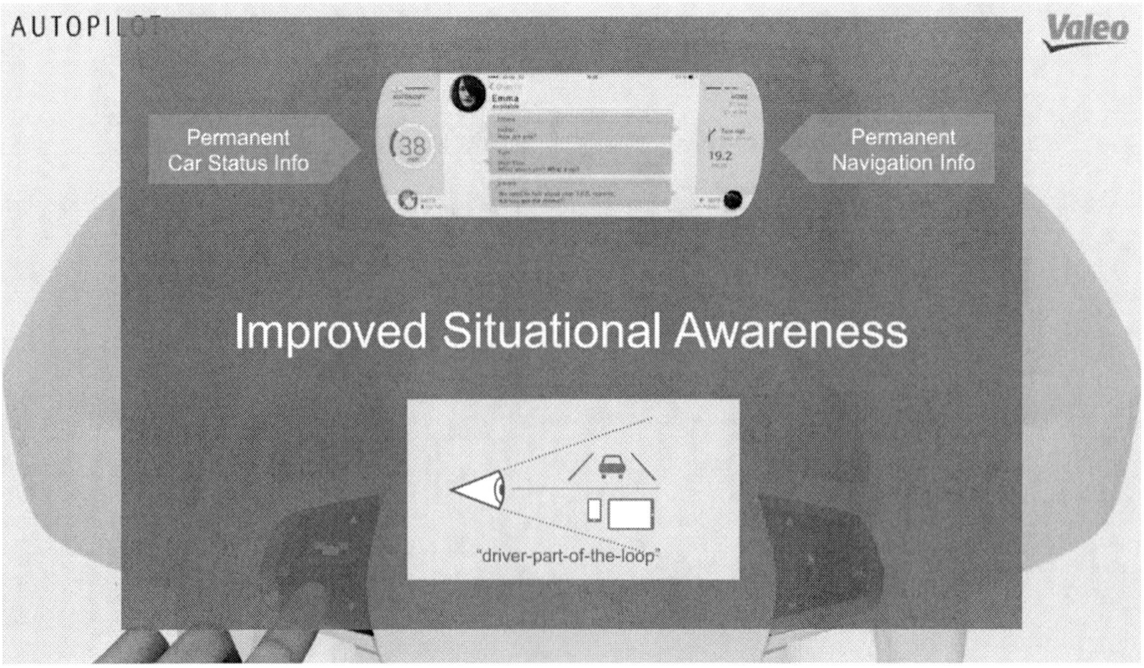
AUTOPILOT
Valeo
Emma
Permanent
Car Status Info
38
19.2
Permanent
Navigation Info
Improved Situational Awareness
"driver-part-of-the-loop"

AUTOPILOT
Valeo
38
19.2
Trust Building & Safer Transition
shorter
unified HMI with driver monitoring

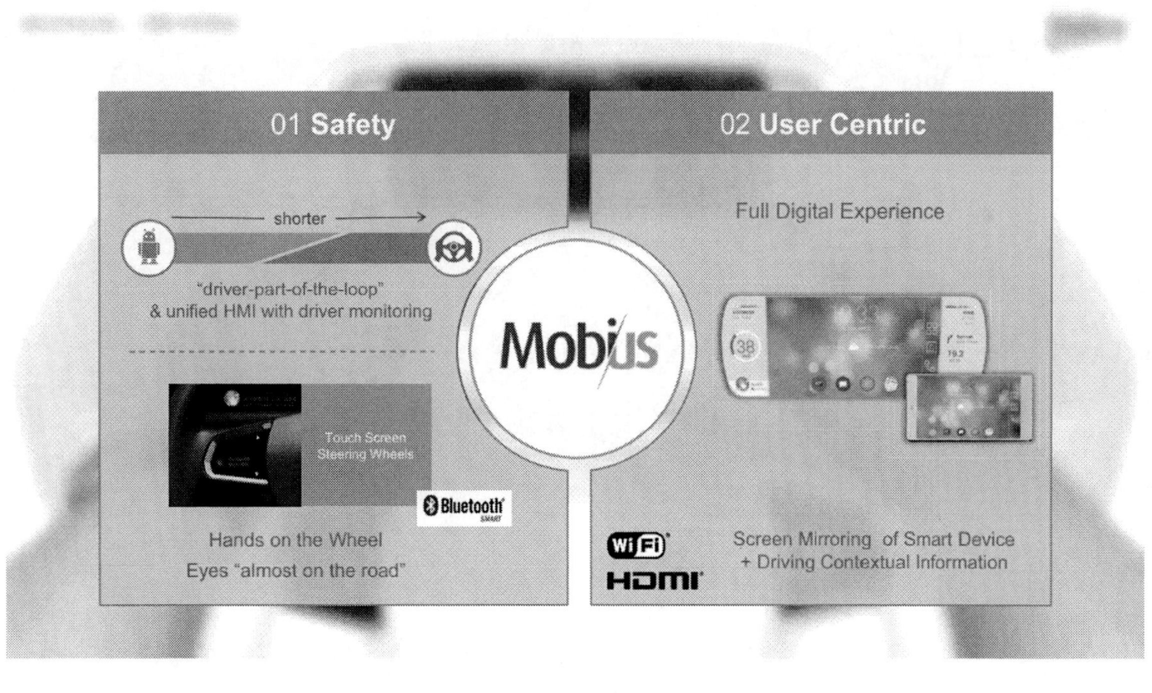
01 Safety
02 User Centric
shorter
"driver-part-of-the-loop"
& unified HMI with driver monitoring
Touch Screen
Steering Wheels
Bluetooth
SMART
Hands on the Wheel
Eyes "almost on the road"
Mobius
Full Digital Experience
38
19.2
WiFi
HDMI
Screen Mirroring of Smart Device
+ Driving Contextual Information

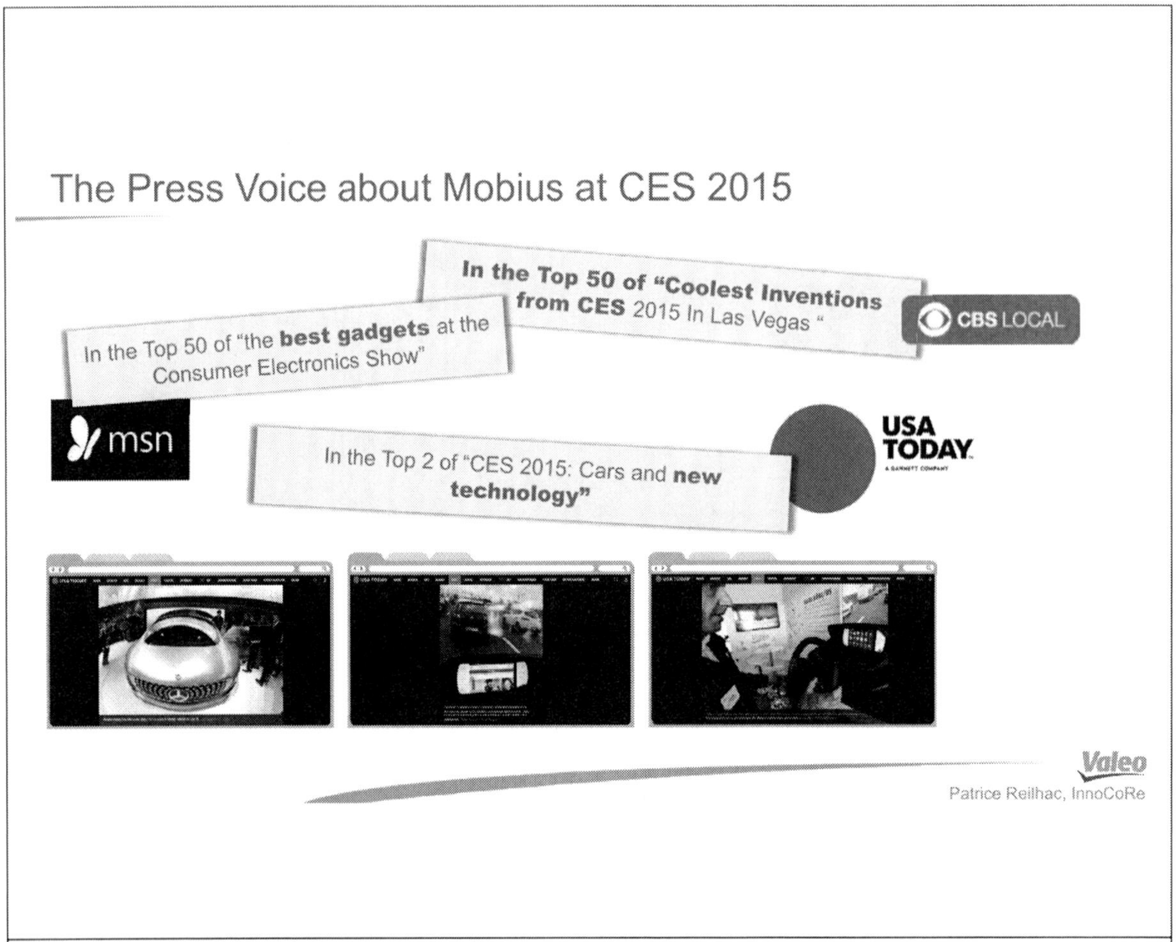
The Press Voice about Mobius at CES 2015
In the Top 50 of "Coolest Inventions from CES 2015 In Las Vegas "
CBS LOCAL
In the Top 50 of "the best gadgets at the Consumer Electronics Show"
msn
In the Top 2 of "CES 2015: Cars and new technology"
USA TODAY
Valeo
Patrice Reilhac, InnoCoRe

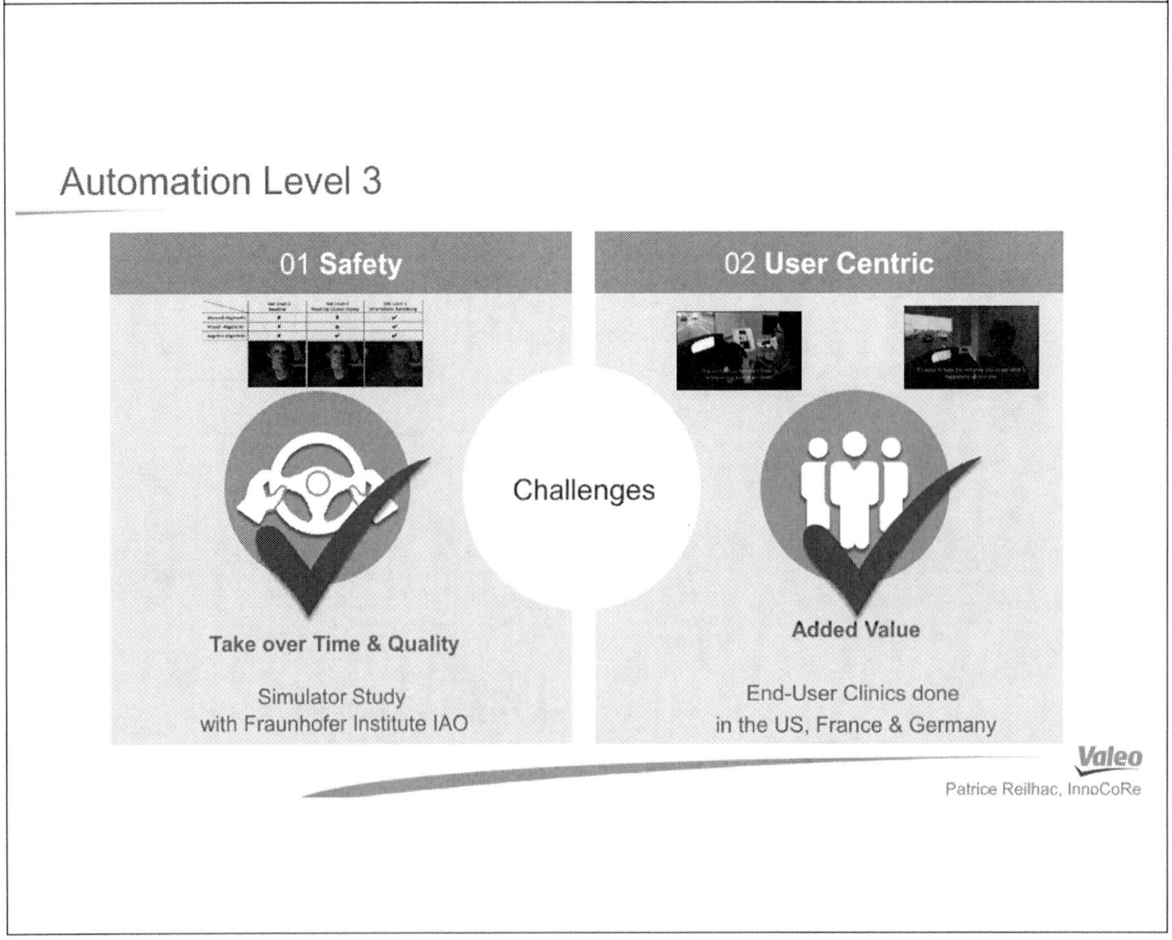
Automation Level 3
01 Safety
02 User Centric
Challenges
Take over Time & Quality
Simulator Study
with Fraunhofer Institute IAO
Added Value
End-User Clinics done
in the US, France & Germany
Valeo
Patrice Reilhac, InnoCoRe

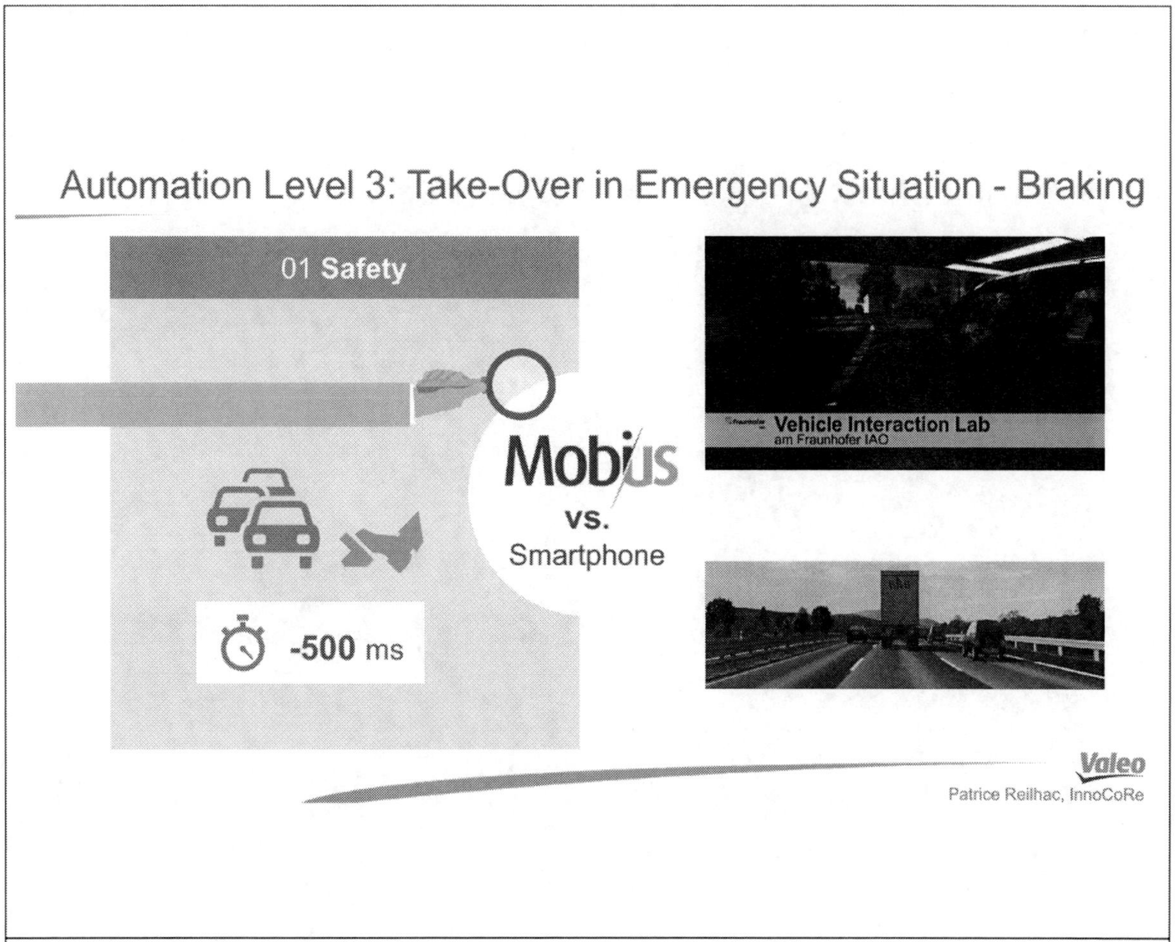
Automation Level 3: Take-Over in Emergency Situation - Braking
01 Safety
-500 ms
Mobius
vs.
Smartphone
Vehicle Interaction Lab
am Fraunhofer IAO
Valeo
Patrice Reilhac, InnoCoRe

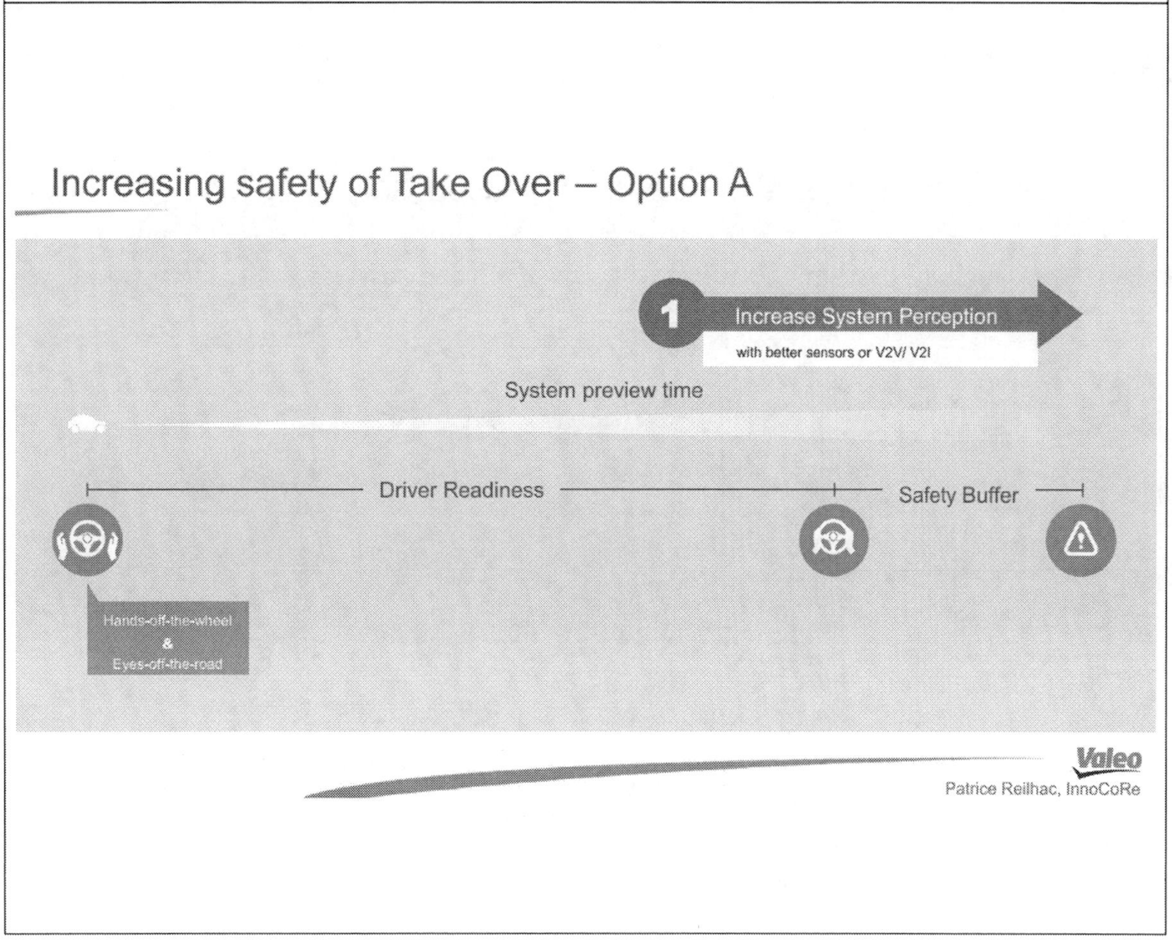
Increasing safety of Take Over – Option A
1
Increase System Perception
with better sensors or V2V/ V2I
System preview time
Driver Readiness
Safety Buffer
Hands-off-the-wheel
&
Eyes-off-the-road
Valeo
Patrice Reilhac, InnoCoRe

Increasing Safety of Take Over - Option B

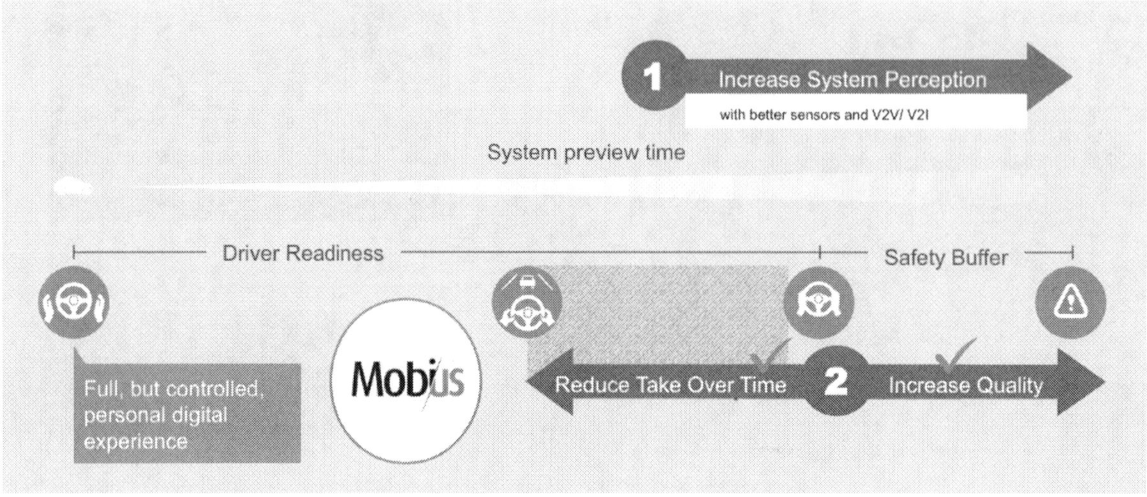

PS: the realworld impact of Mobius on Take Over time & quality needs to be further investigated in simulators & then on the road

Valeo

Patrice Reilhac, InnoCoRe

Need for international collaborative research

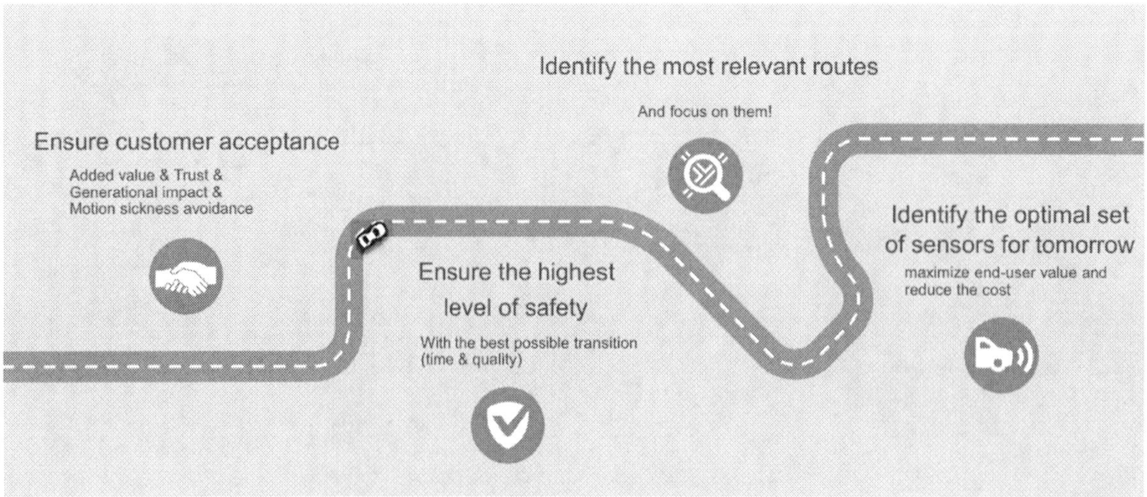

July 15 | 14 Valeo

Next Step: Valeo Mobius 2, UX for automation level 3

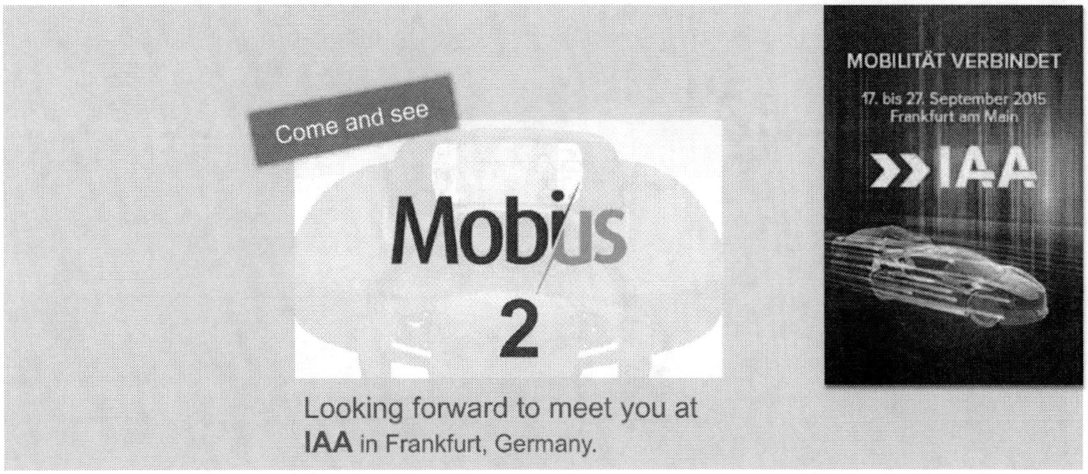

Valeo

Valeo

Automotive technology, naturally

InnoCoRe

Patrice Reilhac
Innovation & Collaborative Research Director
Comfort & Driving Assistance Business Group, Valeo

Connected
Truck Automation

Peloton

US Freight Trucking: $650 Billion

Accident Cost: **$48 Billion**

Fuel Cost: **$100 Billion**

Net Profit: **3%**

Peloton

Connecting Trucks

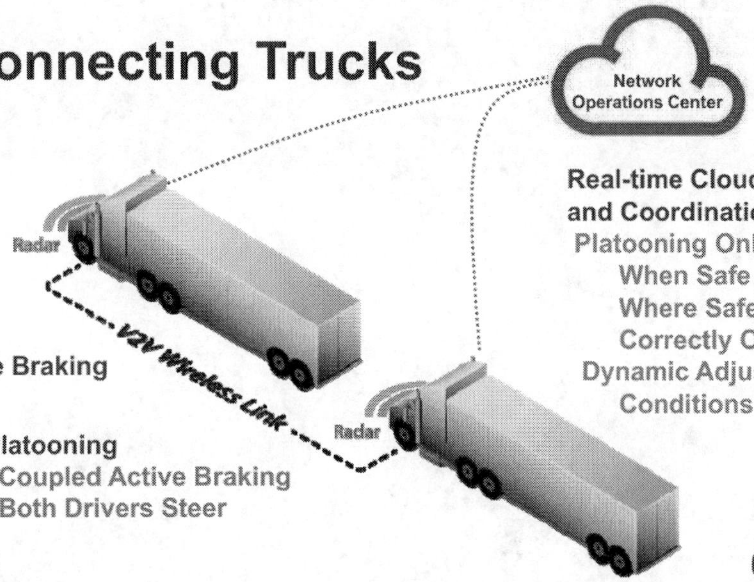

Real-time Cloud Supervision and Coordination
Platooning Only:
When Safe
Where Safe
Correctly Ordered
Dynamic Adjustment to
Conditions

Active Braking

Platooning
Coupled Active Braking
Both Drivers Steer

Peloton

Connected Braking

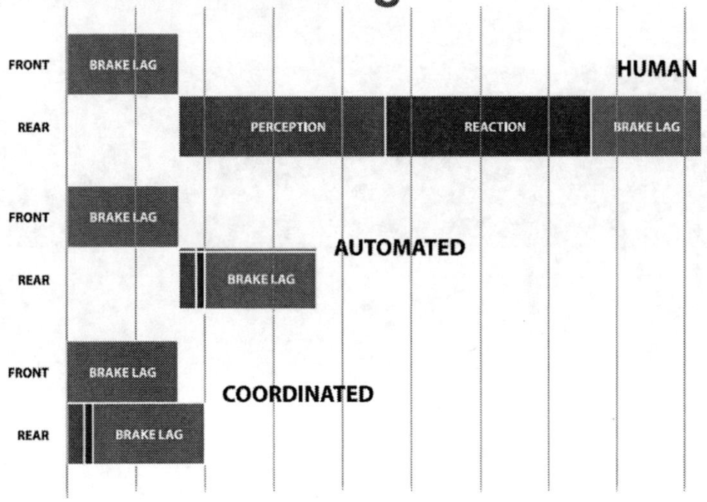

Peloton

Peloton

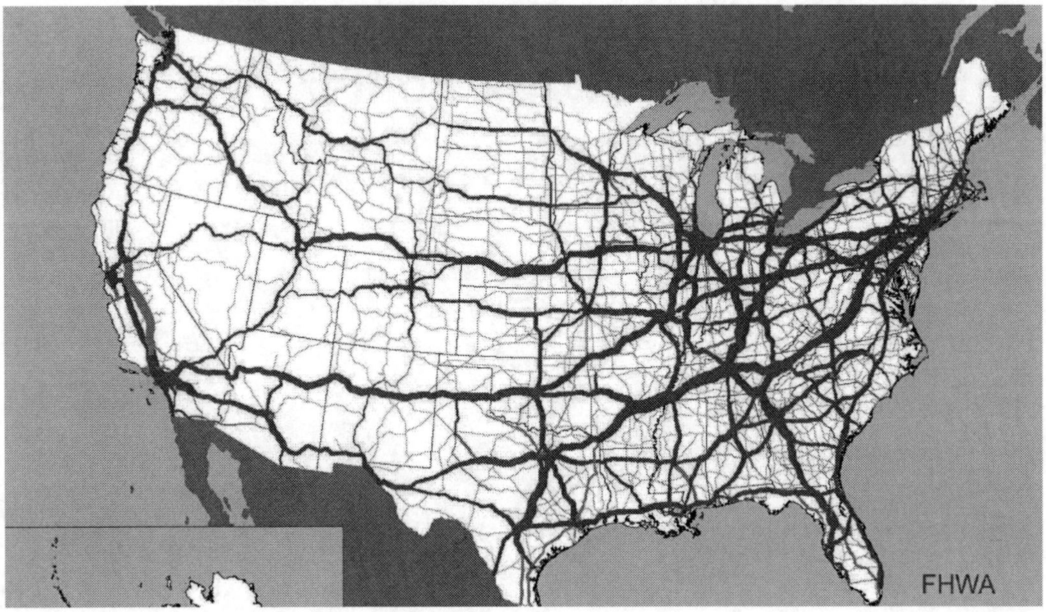
FHWA

Seattle
Washington
Montana
North Dakota
Minnesota
Minneapolis
Quebec
Ottawa
Montreal
Portland
South Dakota
Wisconsin
Vermont
Oregon
Idaho
Wyoming
Milwaukee
Michigan
Toronto
Hamilton
New York
New Hampshire
Massachusetts
Detroit
Chicago
Iowa
Nebraska
Cleveland
Connecticut
New York
Pennsylvania
Ohio
Nevada
Denver
United States
Illinois
Indiana
Philadelphia
Indianapolis
Columbus
Maryland
New Jersey
Sacramento
Utah
Colorado
St. Louis
Cincinnati
West Virginia
Washington
Delaware
San Francisco
Kansas
Missouri
San Jose
California
Kentucky
Virginia
District of Columbia
Las Vegas
Tennessee
North Carolina
Albuquerque
Oklahoma
Arkansas
Charlotte
Los Angeles
Arizona
New Mexico
Atlanta
South Carolina
San Diego
Phoenix
Dallas
Mississippi
Alabama
Georgia
Tucson
Ciudad Juarez
Texas
Baja California
Sonora
Jacksonville
San Antonio
Houston
Louisiana
Chihuahua
Orlando
Tampa
Florida
Coahuila de Zaragoza
Nuevo León
Miami
Baja
Sinaloa
Monterrey

Network Operations Center

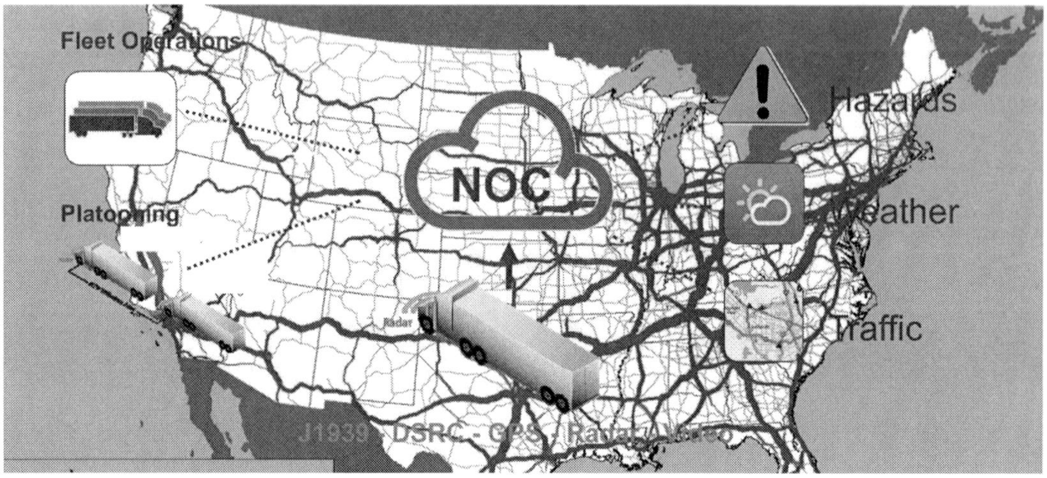

Why Truck Automation First

- Overwhelming economics
- Trained, skilled, experienced drivers
- Highly organized and analytical fleets
- Interstates cover vast majority of freight
- More open architecture

Peloton

Activities

U.S. Department of Transportation

Federal Highway Administration

Peloton

Investors

intel

VOLVO

DENSO

Lytx

NOKIA

ups

bp

MAGNA

LOCKHEED MARTIN

Peloton

Peloton

HERE Automotive

HERE's role in the development of the Digital Transportation Infrastructure

Ogi Redzic
Senior Vice President
HERE Automotive

Agenda

01 HERE Automotive

02 HERE HD Live Map

03 Industry Collaborations

2

HERE Automotive

HD Live Map

Cloud based mapping assets

The HD Map layer is the worlds most precise set of mapping assets enabling precise positioning for lateral & longitudinal vehicle control on the road surface

- 3D Spline-based Surface Model
- ADAS attributes (height, slope, etc)
- Localization objects for positioning

Providing public test track data in US, Europe and Japan and have a planned roll-out for a full-featured HD Live Map product supporting vehicle automation.

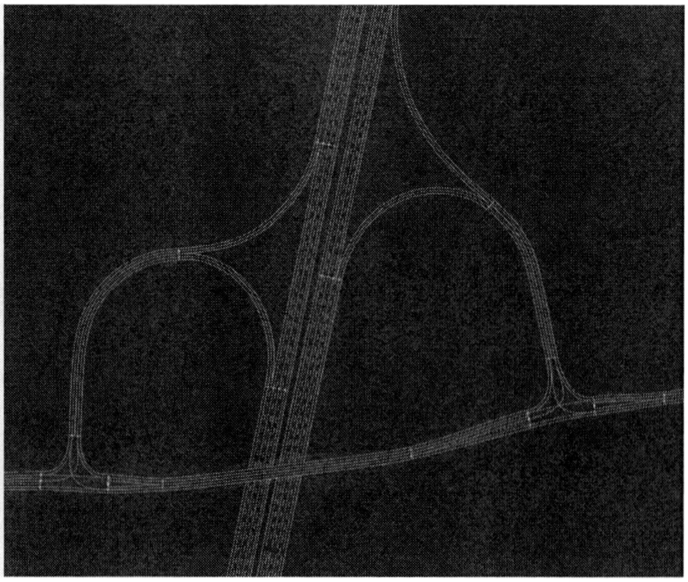

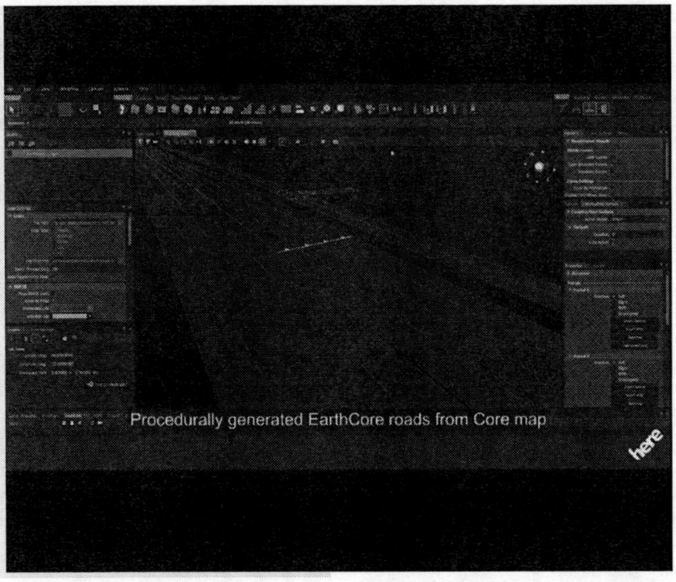

HERE Automotive

How the HD Live Map is built

The worlds most robust data collection and processing practices

Massive investment in collection technology and field deployment with a fleet of over 300 vehicles globally.

Combination of highly precise LiDAR collection technologies fused with OEM vehicle sensor data to create the most precise representation of reality

HERE Automotive

HD Live Map

Real-time data processing

Live Roads is a data aggregation and delivery layer providing real-time awareness, beyond the range of vehicle sensors - allowing insight into:

- congestion
- accidents
- slippery road conditions
- variable message signs

Today HERE collects billions of probe points per day to generate real-time traffic conditions

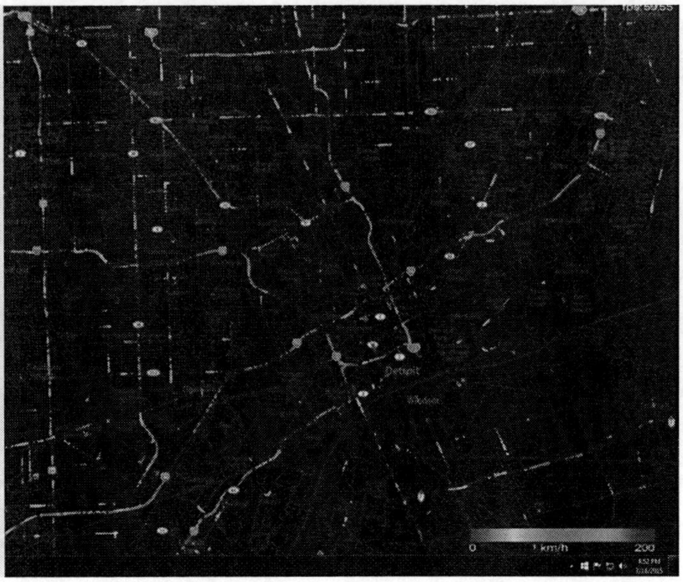

HERE Automotive

HD Live Map

Advanced data analytics

The Humanized Driving layer uses historical probe data to determine speed profiles, with high spatial granularity (10 meters)

Test speed profiles by comparing with "ground truth" collected by HERE drivers all over the world

Clear agreement between median speeds (black) and ground truth (red)

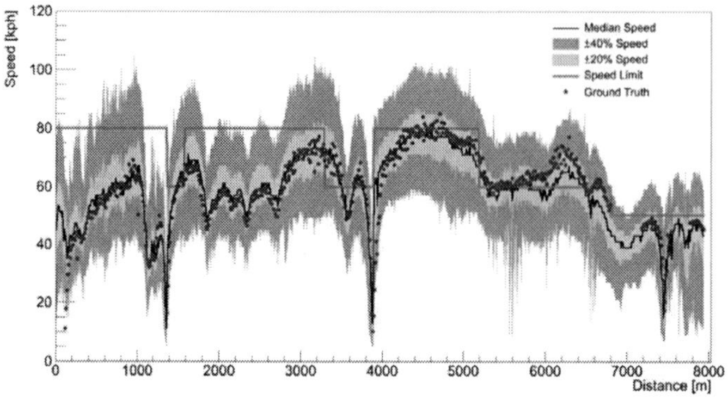

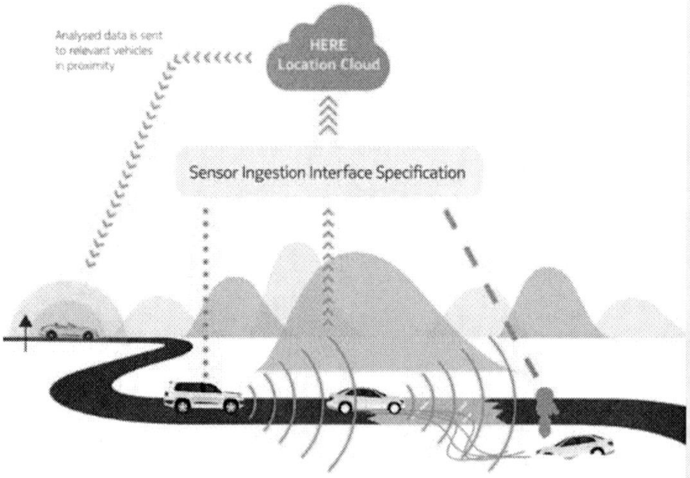

HERE Automotive

Sensor Ingestion Interface Specification

We have announced an open interface specification to allow vehicles to submit rich sensor data back to the cloud.

The goal is to drive this specification to simplify the housekeeping between sensor data suppliers, the cloud and sensor data consumers (vehicles).

HERE Automotive

Michigan Mobility Transformation Center (MTC)

As an industry initiative, MTC was set up with the goal to accelerate connected and automated driving as an industry initiative.

HERE is collaborating with MTC members by providing our HD Live Map for testing piloted automated driving solutions in real world scenarios.

here

HERE Automotive

Finnish Transportation Agency Pilot

First end-to-end integration of C-ITS directive from car to car to traffic management center

Collaboration between government agencies, industry, and research institutions

Hazard warning pilot integrating agency, vehicle, and user data to alert drivers

here

Session Takeaways

Our HD Live Map product is the worlds first cloud based dynamic mapping service powering Automated Driving solutions.

We have announced an open interface specification for sensor data ingestion – fostering industry collaboration for standardizing how vehicles submit rich sensor data back to the cloud.

We are engaged in multiple industry collaboration pilots showcasing our role as the leading location cloud and service provider for smart cities and vehicle automation.

Automated Driving

Automated Vehicles Symposium, July 22, 2015

Ibro Muharemovic, Head of Advanced Engineering NAFTA

Chassis & Safety

SensePlanAct

Not so fun facts

- "Nine out of ten drivers rate their driving skills above average. Correct is, that nine out of ten accidents are caused by human failure." [1]
- "In Germany overtired [sic] drivers in microsleep events are responsible for every fourth traffic death." [2]
- "...stress, aggression and distraction cause driving mistakes and accidents again and again" [3]
- "In the U.S., distraction contributes to approximately 80% of crashes and 16% of fatalities" [4]

The Phenomenon

A "Head Down" Society

Continental at a Glance

More than 140 Years of Progress and Achievement

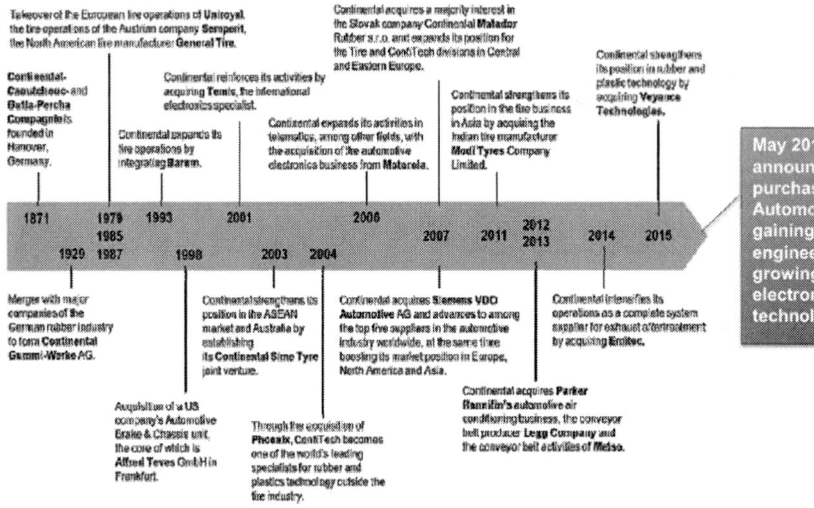

Software Challenge

Paramount Importance to keep the pace!

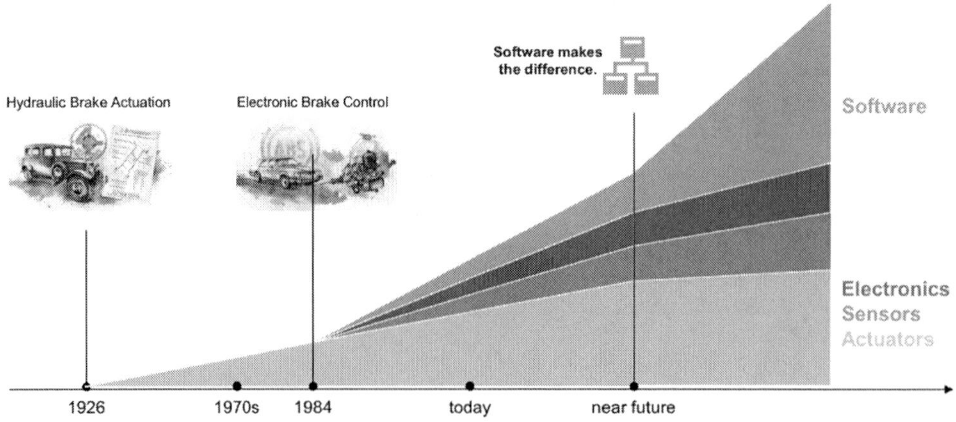

The Three Pillars of Automated Driving

More safety, comfort and efficiency

Safety
- Less accidents
- Convenient ride

Comfort
- Personal mobility
- Private space
- Freedom of time utilization

Efficiency
- Smooth traffic flow, less fuel consumption

The Safety Basics

If a car assists you, or drives you automatically, it must

Identify & Interpret | **Decide & Control** | **Execute**

Safety
The Basis for the Introduction of Advanced Technologies

Automated Driving @ Continental
A Strong History

DARPA Urban Challenge 2007

› Autonomous Driving with some Continental Components and Embedded Engineer

Research partner and project lead in HAVEit

› EU funded project on automated driving
› First function demonstration with sensors from series production

Nevada AD driving license in 2012

› Driving on public roads

Learning through Highly Automated Driving
Continental HAD on-the-road since 2012

Safety
The Basis for the Introduction of Advanced Technologies

New Vehicle Architecture with Redundancies and Fallbacks:

- Power Supply
- Communication
- Sensing
- Computing
- Braking
- Steering

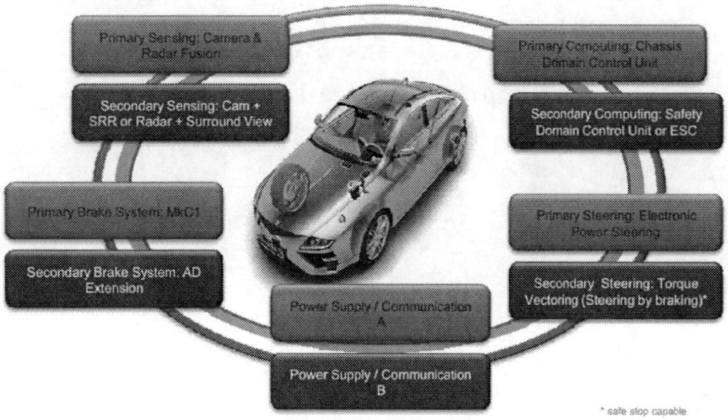

Thank you

References

- [1] Interview, Daimler-Vorstandsvorsitzender Dieter Zetsche, AutoMotorsport, Heft 24
- [2] ADAC e.V.: Müdigkeit im Straßenverkehr – unterschätzt, verkannt, tödlich. Artikelnummer 2831141, www.adac.de/infotestrat/ratgeber-verkehr
- [3] Elektronik automotive 10.2014, "Fahrerzustandsmodellierung im Fokus", von Alexander König, Ralf Decke, und Ralph H. Rasshofer.
- [4] ww.aaafoundation.org

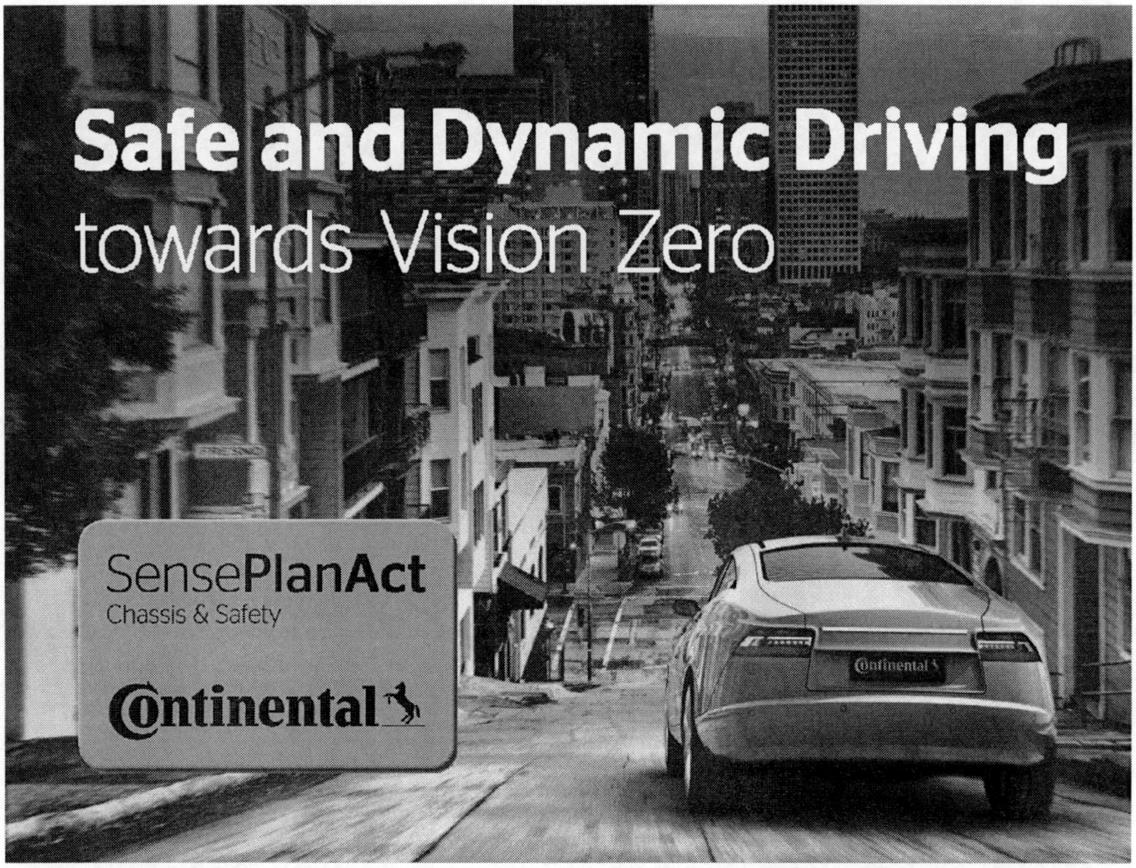
Safe and Dynamic Driving
towards Vision Zero
SensePlanAct
Chassis & Safety
Continental

INSTITUT FÜR KRAFTFAHRZEUGE
ika | RWTH AACHEN UNIVERSITY

Adapt!Ve

Automated Driving Applications and Technologies for Intelligent Vehicles

Felix Fahrenkrog
Institut für Kraftfahrzeuge
RWTH Aachen University

AdaptIVe - Evaluation

Automated Vehicle Symposium 2015

Ann Arbor
July 22nd, 2015

// AdaptIVe

//Validation & Verification

Definition according to "IEEE Guide--Adoption of the Project Management Institute Standard A Guide to the Project Management Body of Knowledge"

- Validation
  - The assurance that a product, service, or system meets the needs of the customer and other identified stakeholders. It often involves acceptance and suitability with external customers.
- Verification
  - The evaluation of whether or not a product, service, or system complies with a regulation, requirement, specification, or imposed condition. It is often an internal process.

- Main objectives of SP7 in AdaptIVe:
  - Evaluation of automated driving research functions in four assessments (technical, user-related, in-traffic, impact assessment)
  - Development of an evaluation framework for automate driving functions
    - **Verification** is part of the **technical assessment** (→ checking requirements)
  - Methodology for impact analysis of automated driving applications
    - **Validation** is part of safety and environmental **impact assessment** as well as the user-related, in-traffic and **technical assessment**

//Classification automated driving functions

- Classification of automated driving functions for the evaluation
  - According to the SAE definition
  - According the operation time
    - **Event based operating**
      - Function that operates for a short period in time (typically vehicle stands still at the end or the automated driving ends)
    - **Continuously operating**
      - Function that operates for a longer period in time (typically vehicle is still moving at the end of an manoeuvre respectively automated driving is continued)

//Evaluation approach technical assessment

- How to limit the test amount to feasible amount?

Different approaches for event-based and continuously operation function

- Event-based functions: similar approaches as in previous research project e.g. interactIVe
- Continuously operating functions: test on public road

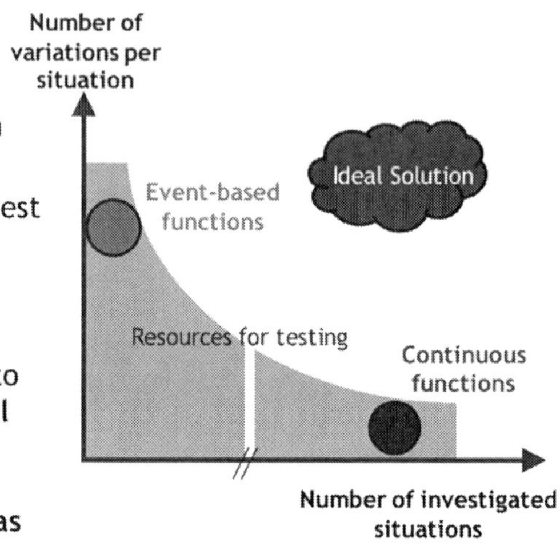

Evaluation criteria

- Automated driving functions need to operate within the **range of normal driving behaviour** (→ not disturb normal driving in mixed traffic conditions) and should **at least be as safe** as non-automated driving

//Evaluation approach technical assessment

//

1. **Defining evaluation scope**
   - Definition of research questions, hypotheses & indicators
2. **Planning of assessment**
   - Analyse system description and adaption of hypotheses
   - Planning of test cases
   - (Risk assessment)
3. **Tests in controlled field**
   - Number of test variations
   - Logging of test data
4. **Assessment of tests**
   - Analysis of hypotheses based on test data & indicators

//

1. **Defining evaluation scope**
   - Definition of research questions, hypotheses & indicators
2. **Planning of assessment**
   - Analyse system description and adaption of hypotheses
   - Planning of test cases and test route
   - Definition evaluation criteria (distributions & boundaries)
   - Risk assessment
3. **Pre-/component tests in controlled field**
   - Basic tests of functionality
   - Sensor tests
4. **Tests in real traffic**
   - Test route and test amount to be determined
5. **Assessment of tests**
   - Analysis of hypotheses based on test data & indicators

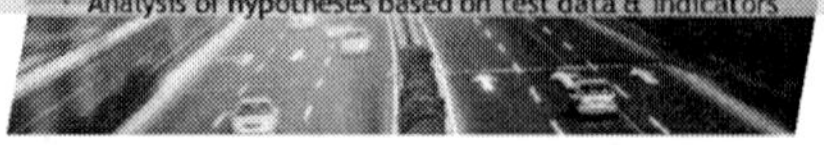

Simulation (SiL)
- In case test effort for other test tools is too high or tests are not feasible (optional)
- Higher relevance for the (serial) development

// Safety Impact Assessment

- Research Question:
  - What is the safety benefit of automated driving functions?
- State of the art approaches for ADAS:
  - Field of application
  - Accident re-simulation
  - Field test
- Open issues for the impact assessment of automated driving applications
  - Today's accident data do not consider collisions of automated vehicles
  - Automated driving function operate already before a critical situation occurs → Re-simulation of accidents gets more difficult
  - Interaction with other road users (automated / non-automated) → mixed traffic

Source: http://files.conceptcarz.com, www.7-forum.com, ika

// Safety Impact Assessment

Safety impact assessment in AdaptIVe bases on three main steps

1. **Identify relevant situations**
   - Focus on accident and other (relevant) driving situations
   - Use of microscope traffic simulation
2. **Investigate the relevant situation in detail**
   - Approach is similar to the re-simulation approach
   - Input data from other assessment (technical, user-related, in-traffic assessment) are considered
3. **Identification of new situations**
   - e.g. transition of control or minimum risk maneuver

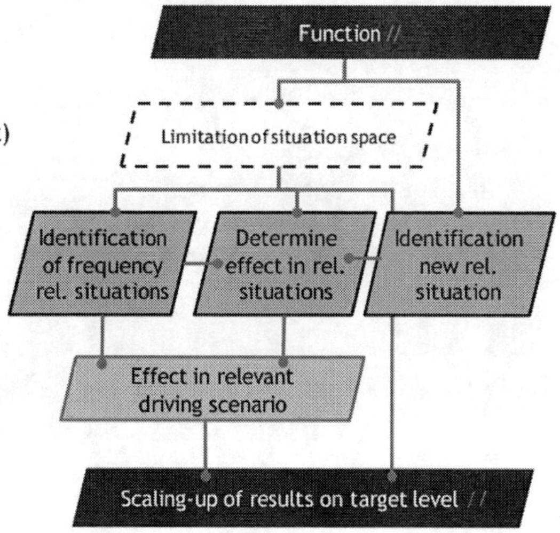

//Outlook: ika's approach for the safety validation

INSTITUT FÜR KRAFTFAHRZEUGE RWTH AACHEN UNIVERSITY

- How to validate / verify that the a automated driving functions safe enough for the market introduction?
  - Circle of relevant situation approach [ECK13] [ZLO15]

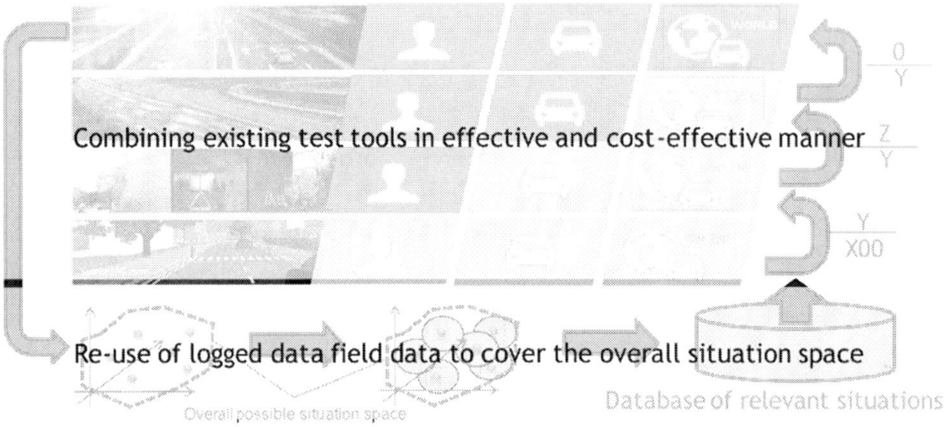

[ECK13]: Eckstein, Zlocki; Safety Potential of ADAS - Combined Methods for an Effective Evaluation; 23rd ESV; 2013
[ZLO15]: Zlocki, Eckstein, Fahrenkrog; Evaluation and sign-off methodology for automated vehicle systems based on relevant driving situations; 94th Annual TRB Meeting; Washington D.C.; 2015

9 // 22 July 2015 Automated Vehicle Symposium, Ann Arbor *Adapt!Ve*

Co-funded by the European Union

INSTITUT FÜR KRAFTFAHRZEUGE RWTH AACHEN UNIVERSITY

Adapt!Ve

Automated Driving Applications and Technologies for Intelligent Vehicles

Thank you.

Felix Fahrenkrog
Institut für Kraftfahrzeuge,
RWTH Aachen University
fahrenkrog@ika.rwth-aachen.de

Automated Vehicle Symposium 2015
Public Agency Automated Vehicle Initiative

SIP-adus: a National Project in Japan

Cross-Ministerial Strategic Innovation Promotion Program
Innovation of Automated Driving for Universal Services

July 23, 2015

Hajime Amano
Chair, International Corporation WG, SIP-adus
President, ITS Japan

SIP

National Strategies by Abe Administration

Three Arrows of the Economic Policies

1st Bold Monetary Policy

2nd Flexible Fiscal Policy

3rd **Japan Revitalization Strategy** (New Growth Strategy)

Closely Linked

Science and Technology Policies

Science, Technology and Innovation Strategy

Council for Science, Technology and Innovation Policy (CSTI)

SIP

1

Cross-Ministerial Strategic Innovation Promotion Program

CSTI

Governing Board

Program Director

Promoting committee
- Program Director (chair)
- Related ministries
- Management agencies
- Experts from academia and public sector

Funding Agency

Research organizations
- Universities,
- Private companies,
- Research institutes, etc.

Established for each project

Automated Driving System

Program Director
Dr. Hiroyuki Watanabe
Advisor,
Toyota Motor Corporation

2

Portfolio of ITS deployment since 1996

1. Car Navigation

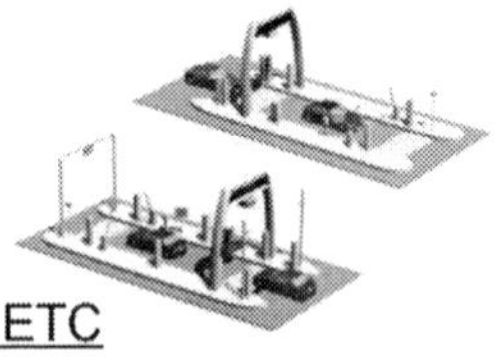

2. ETC

3. Driving Safety

4. Traffic Management

5. Road Management

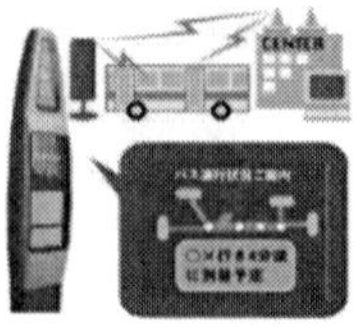

6. Public Transportation

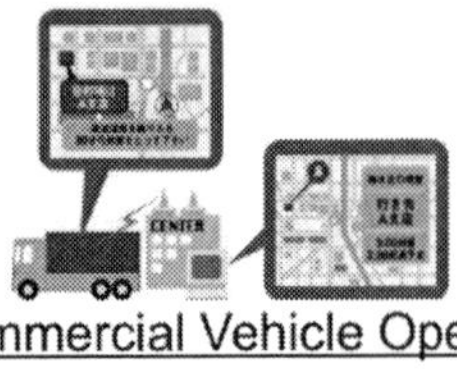

7. Commercial Vehicle Operation

8. Pedestrian Support

9. Emergency Vehicle Operation

3

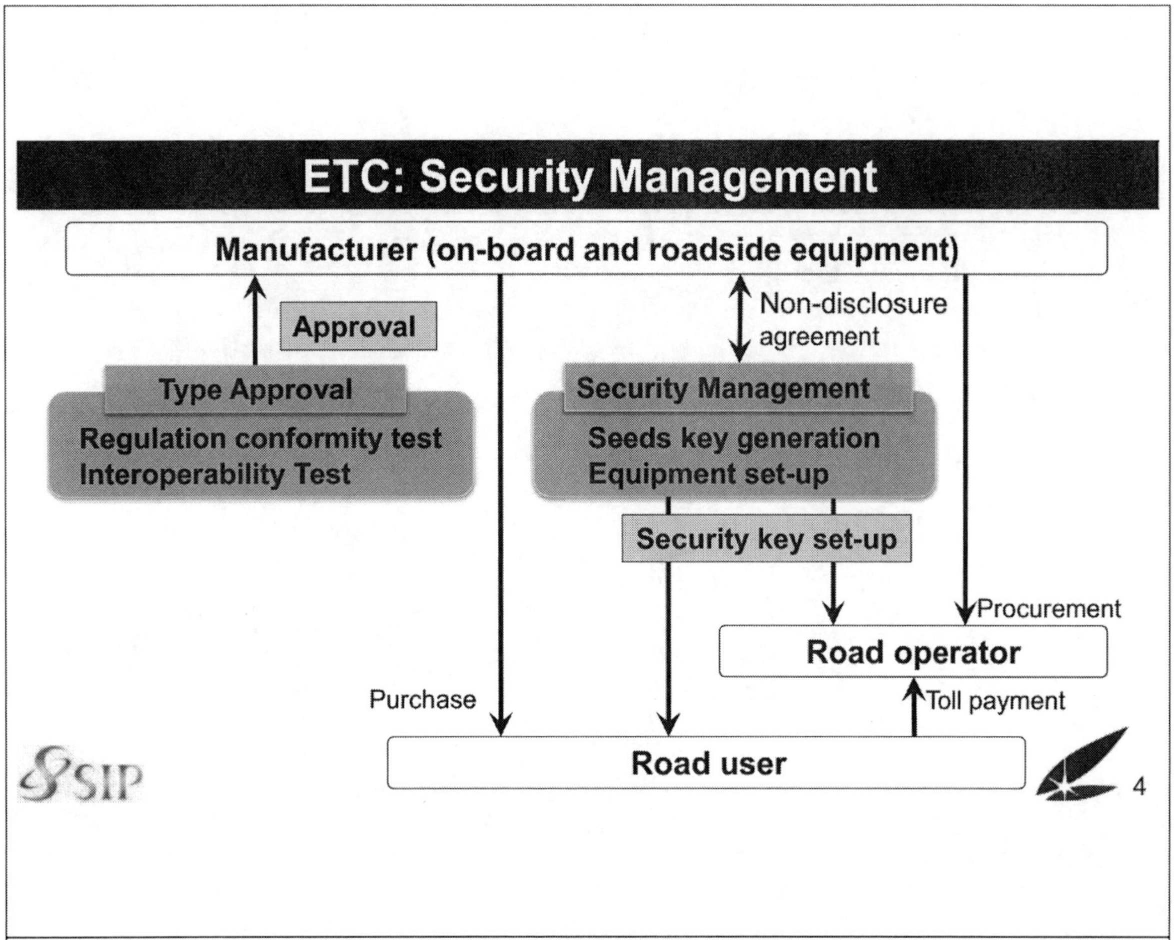
ETC: Security Management
Manufacturer (on-board and roadside equipment)
Approval
Non-disclosure agreement
Type Approval
Regulation conformity test
Interoperability Test
Security Management
Seeds key generation
Equipment set-up
Security key set-up
Procurement
Road operator
Purchase
Toll payment
Road user
SIP
4

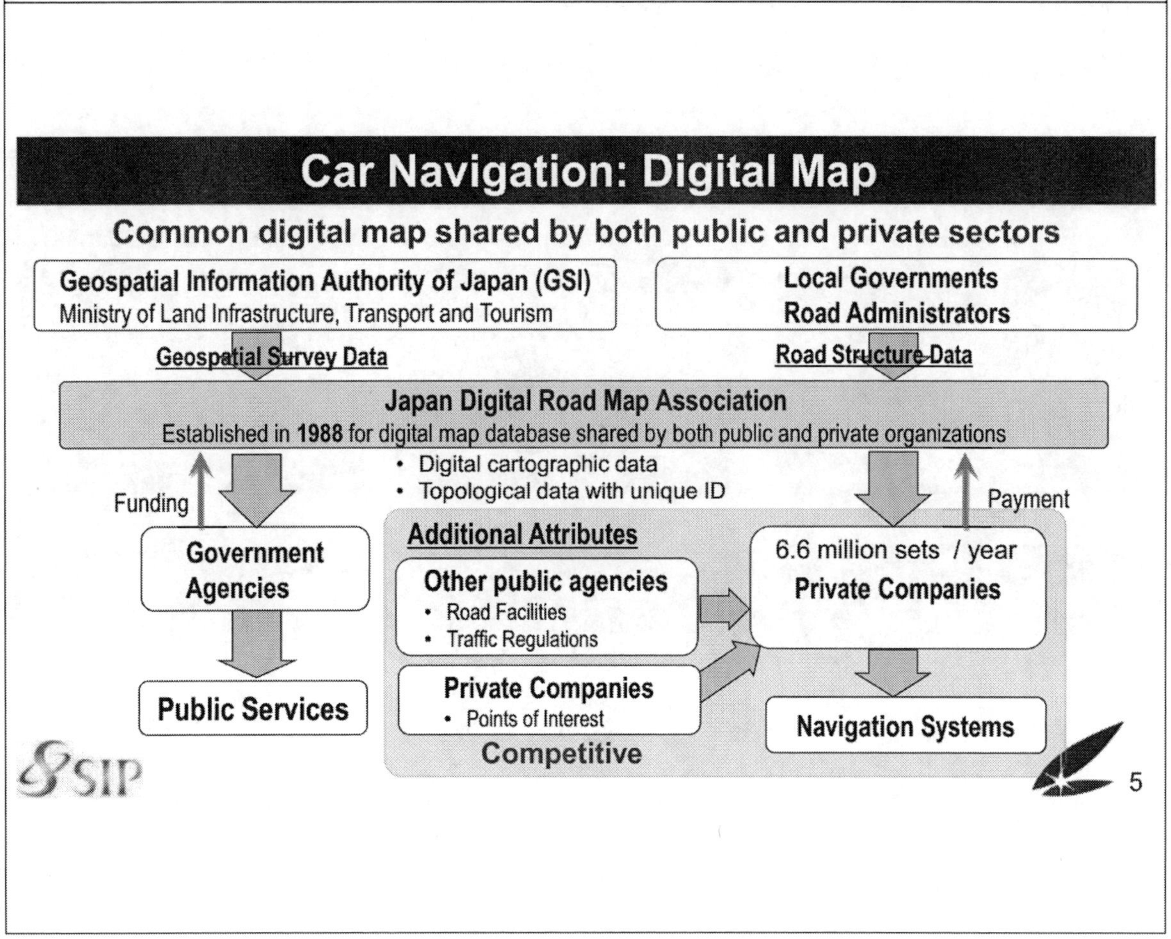
Car Navigation: Digital Map
Common digital map shared by both public and private sectors
Geospatial Information Authority of Japan (GSI)
Ministry of Land Infrastructure, Transport and Tourism
Local Governments
Road Administrators
Geospatial Survey Data
Road Structure Data
Japan Digital Road Map Association
Established in 1988 for digital map database shared by both public and private organizations
• Digital cartographic data
• Topological data with unique ID
Funding
Payment
Government Agencies
Additional Attributes
6.6 million sets / year
Private Companies
Other public agencies
• Road Facilities
• Traffic Regulations
Private Companies
• Points of Interest
Competitive
Public Services
Navigation Systems
SIP
5

VICS: Common Reference on Any Map

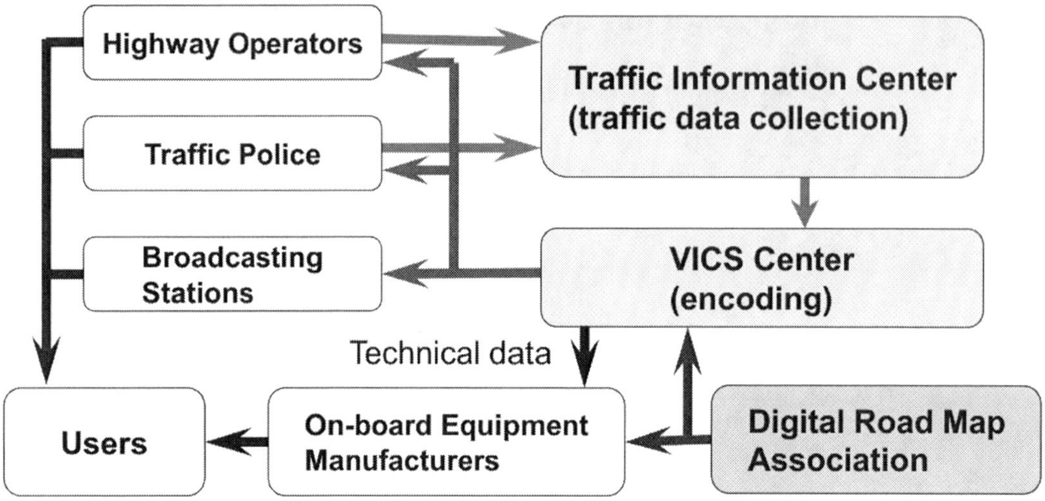

Common digital map is shared by both public and private sectors

6

Cooperative Services: In Operation Since 2011

Safety Assistance | Traffic Information | Dynamic Route Guidance

7

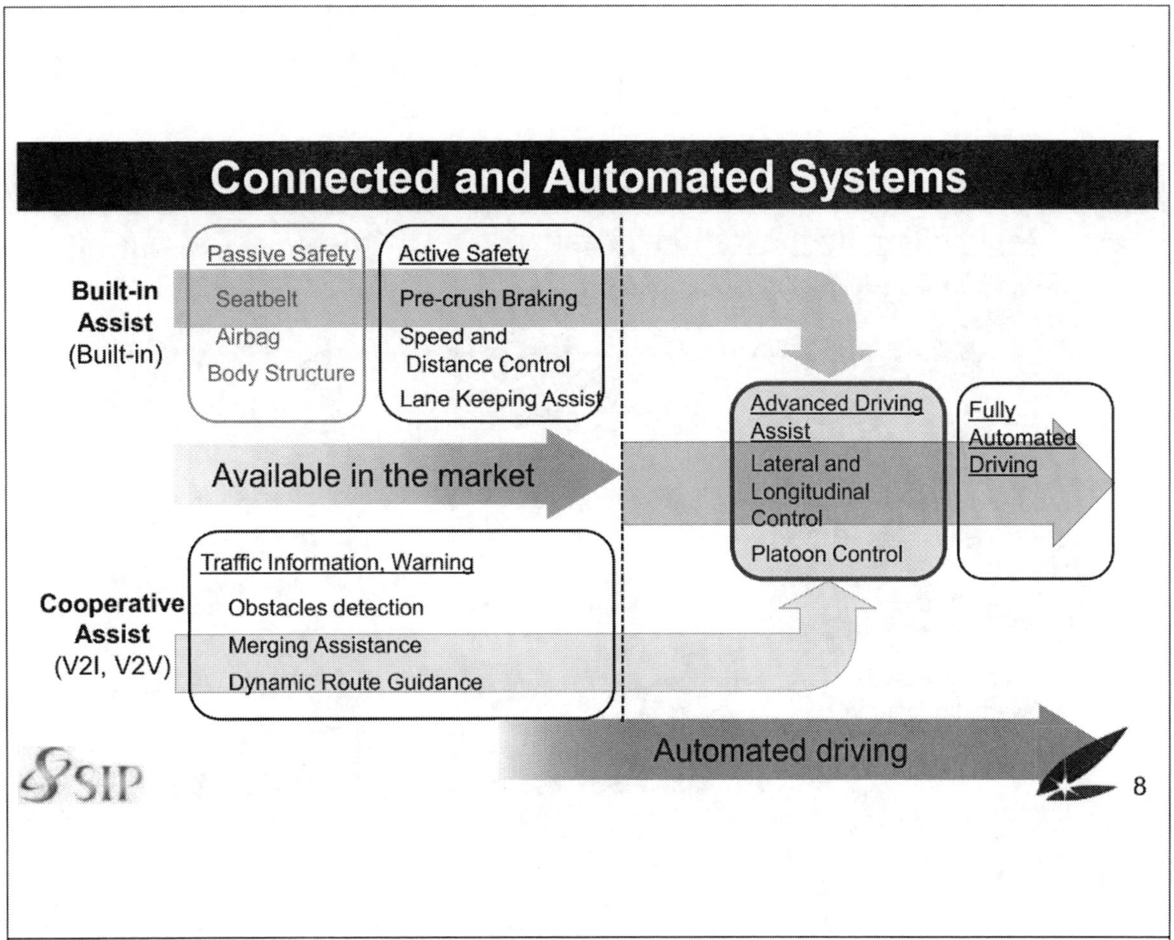
Connected and Automated Systems
Passive Safety
Seatbelt
Airbag
Body Structure
Active Safety
Pre-crush Braking
Speed and Distance Control
Lane Keeping Assist
Built-in Assist (Built-in)
Available in the market
Advanced Driving Assist
Lateral and Longitudinal Control
Platoon Control
Fully Automated Driving
Cooperative Assist (V2I, V2V)
Traffic Information, Warning
Obstacles detection
Merging Assistance
Dynamic Route Guidance
Automated driving
SIP
8

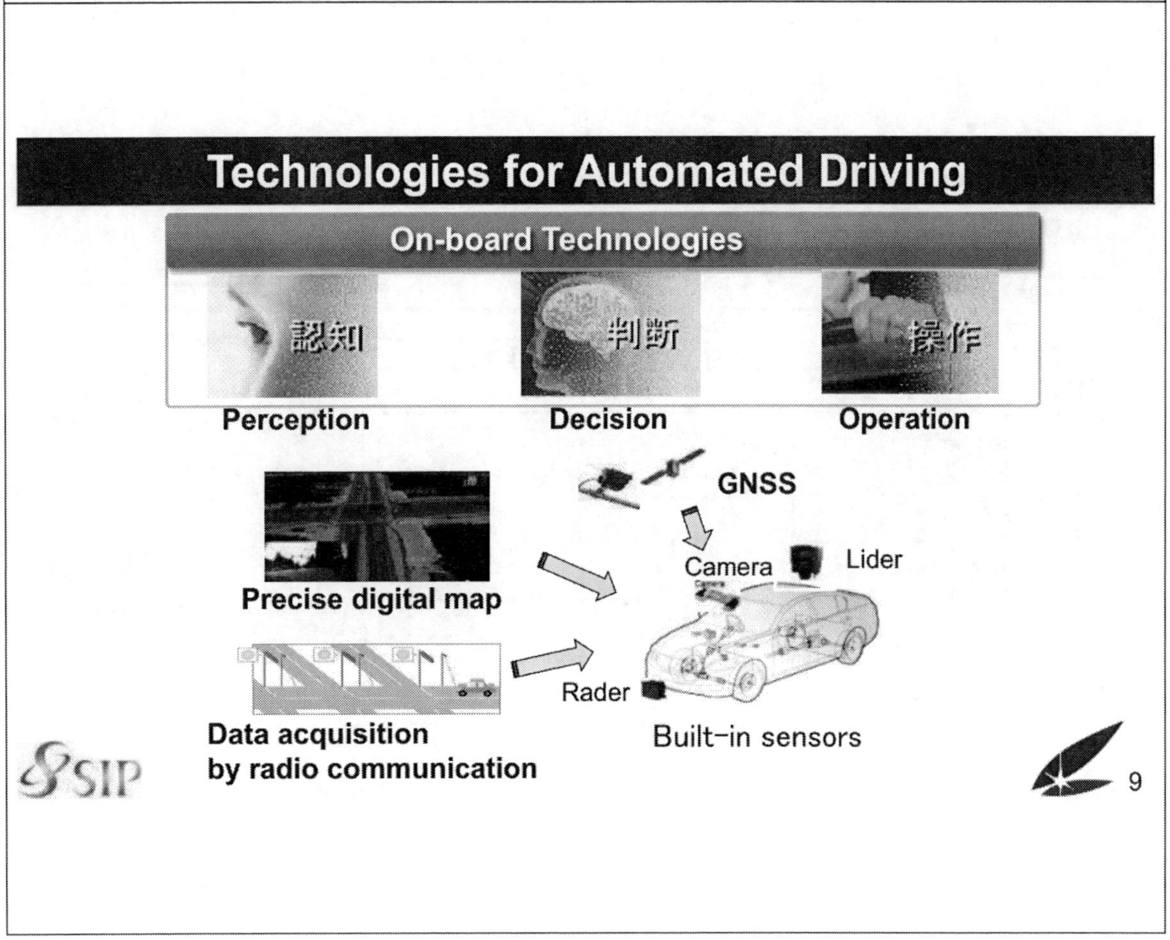
Technologies for Automated Driving
On-board Technologies
認知
判断
操作
Perception
Decision
Operation
GNSS
Camera
Lider
Precise digital map
Rader
Data acquisition by radio communication
Built-in sensors
SIP
9

Auto Manufacturers already competing

Recognition by the National Leader

Prime Minister Abe showed strong interest.

Showcase in Detroit

SIP

10

Scope of SIP-adus

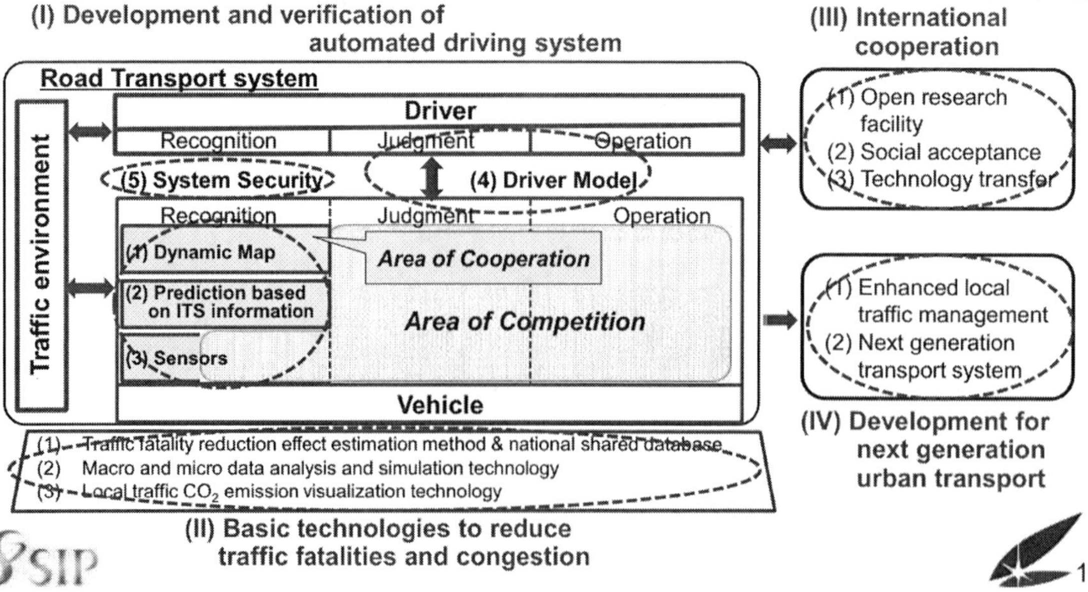

SIP

11

Dynamic Map

Hierarchical structure of digital 'Map' layered by time frame

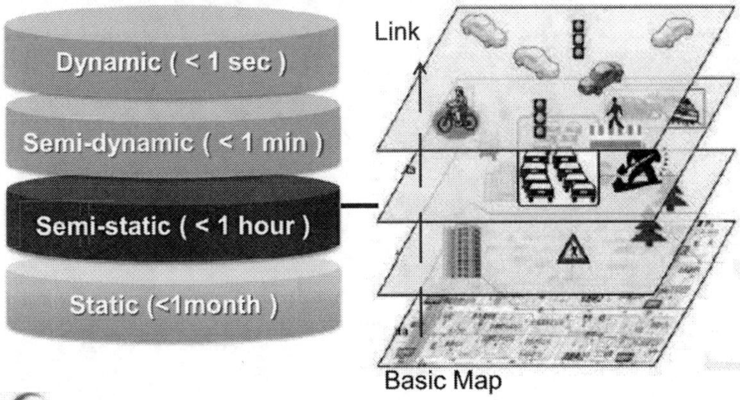

Information through V to X
- surrounding vehicles
- pedestrians
- timing of traffic signals

Traffic Information
- accidents
- congestion
- local weather

Planned and forecast
- traffic regulations
- road works
- weather forecast

Basic Map Database
- Digital cartographic data
- Topological data with unique
- Road Facilities

SIP

12

Prototyping Dynamic Map

Road Environment

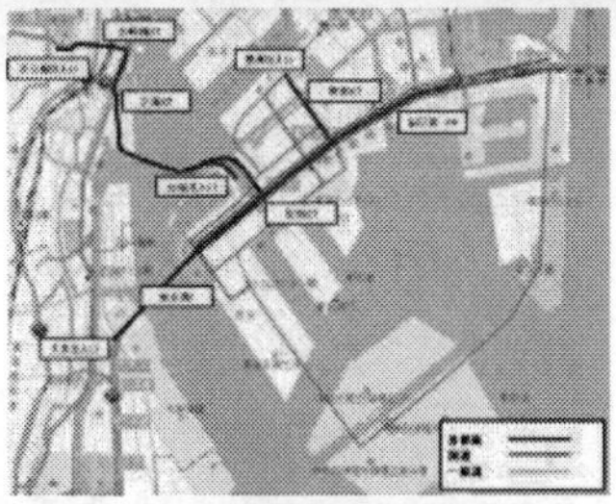

Target Area

3D Measurement

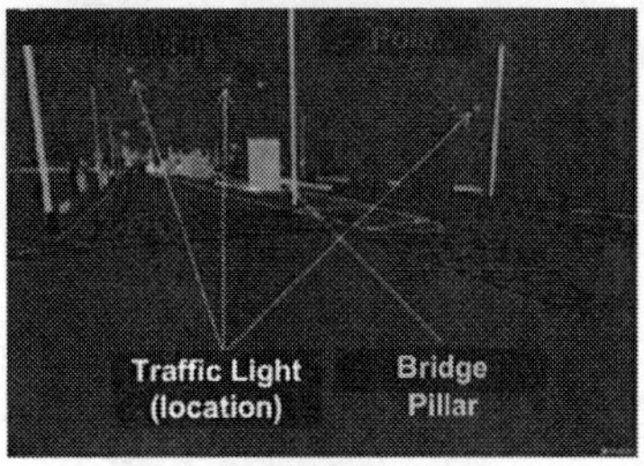

Linked Objects

SIP

13

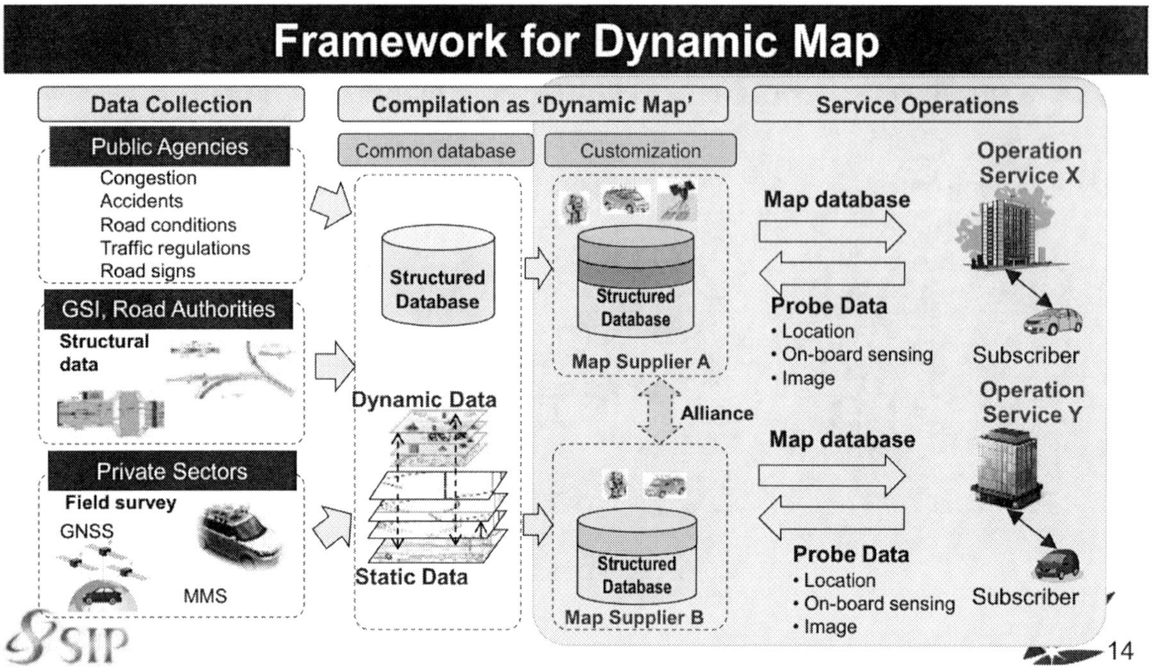
Framework for Dynamic Map
Data Collection
Compilation as 'Dynamic Map'
Service Operations
Public Agencies
Congestion
Accidents
Road conditions
Traffic regulations
Road signs
GSI, Road Authorities
Structural data
Private Sectors
Field survey
GNSS
MMS
Common database
Customization
Structured Database
Dynamic Data
Static Data
Structured Database
Map Supplier A
Alliance
Structured Database
Map Supplier B
Map database
Probe Data
• Location
• On-board sensing
• Image
Map database
Probe Data
• Location
• On-board sensing
• Image
Operation Service X
Subscriber
Operation Service Y
Subscriber
SIP
14

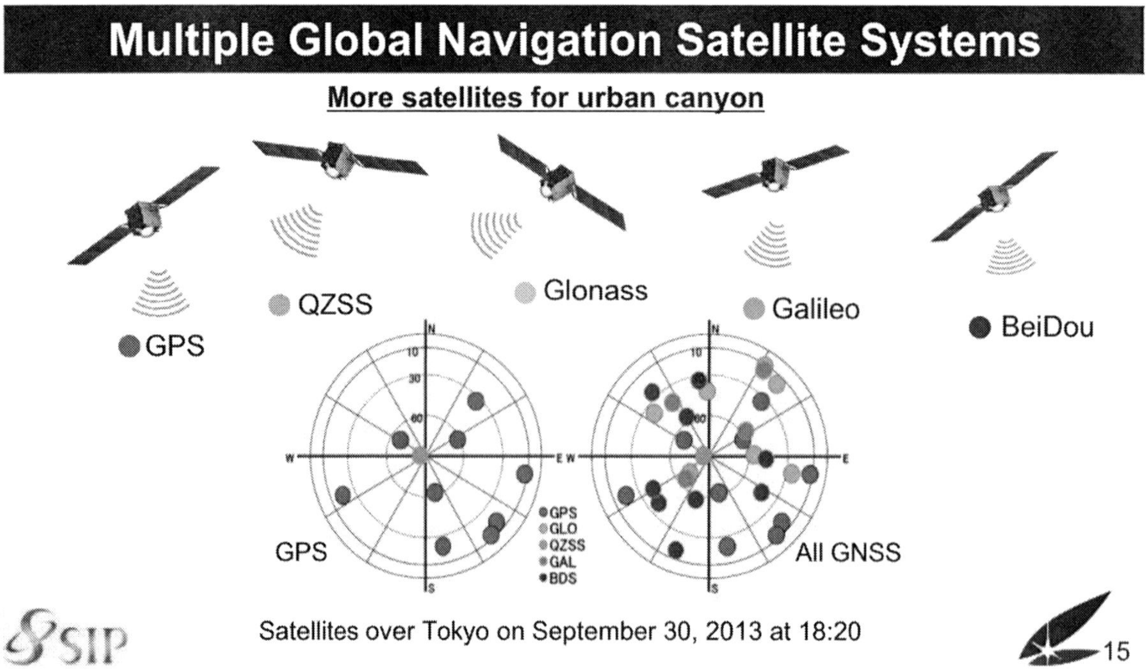
Multiple Global Navigation Satellite Systems
More satellites for urban canyon
GPS
QZSS
Glonass
Galileo
BeiDou
GPS
All GNSS
GPS
GLO
QZSS
GAL
BDS
Satellites over Tokyo on September 30, 2013 at 18:20
SIP
15

Multiple GNSS and Accuracy Enhancement

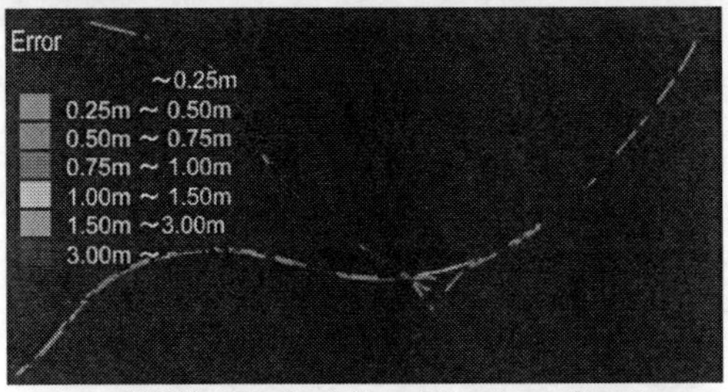

GPS: Global Positioning System

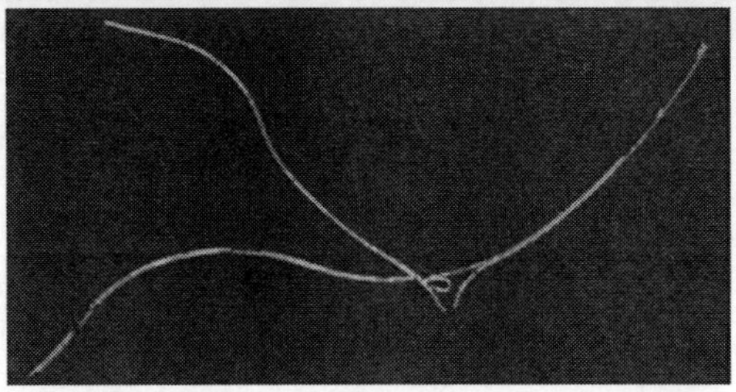

GPS + QZSS with (L1-SAIF) enhancement

GPS + QZSS with (CMAS) enhancement

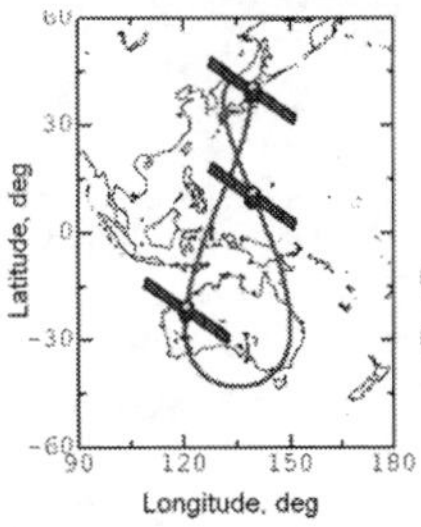

QZSS:
Quasi- Zenith Satellite System

16

Human Factors

Transition of role and authority

- NHTSA Level 3
- Conditional Automation

Transition event

- NHTSA Level 2
- Partial Automation

Human-out-of-the-loop / -in-the-loop

Human-in-the-loop

Use case

Scenario and Interaction

Role	Authority
Function	Condition
Task	Status

! acquisition — Cognition — Judgment — Control

Transition event

SIP

17

Benefits

- **Road traffic safety**
- **Environmental sustainability**
- **Universal transportation services**

with special attention to aging society

SIP

18

Road Traffic Fatality by Age Group

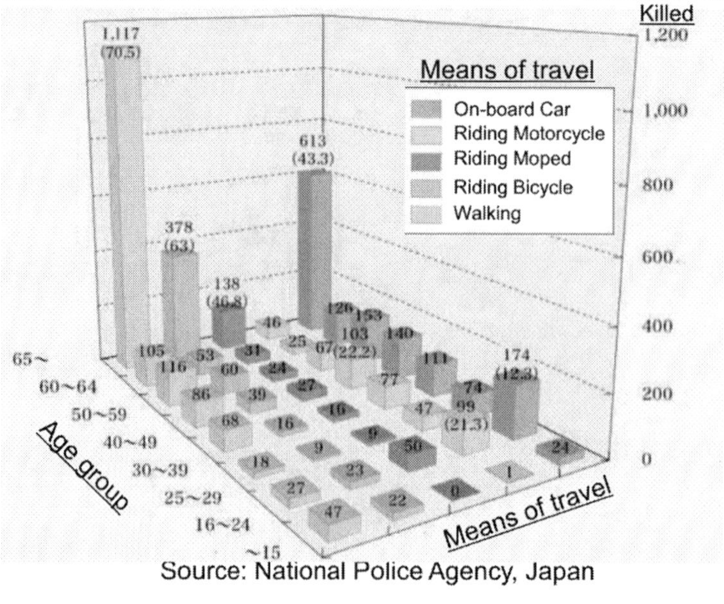

Source: National Police Agency, Japan

SIP

19

Fatal Crash caused by Violation

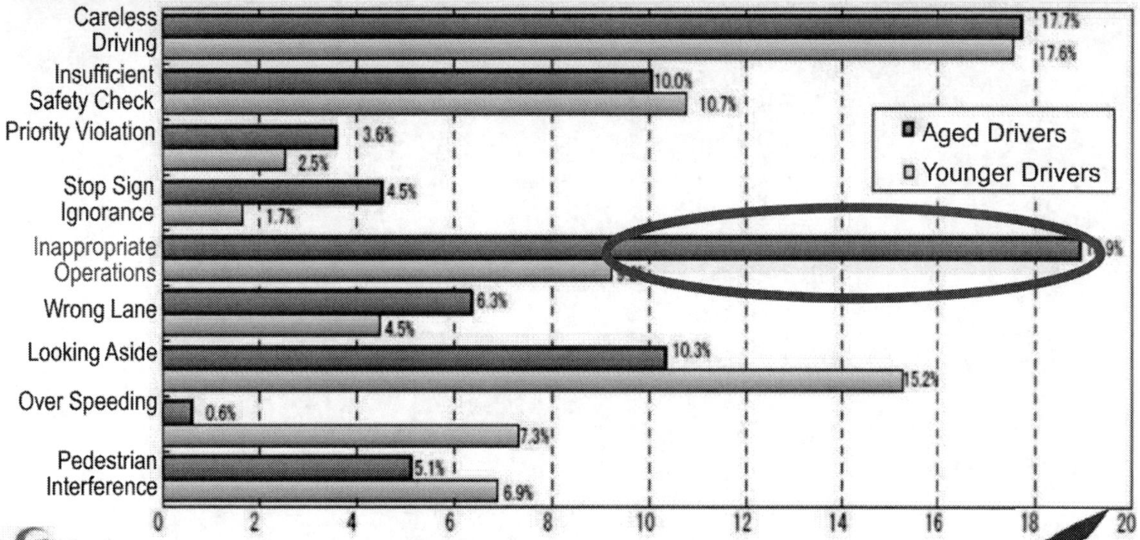

Source: National Police Agency, Japan

Demands from Families of Victims

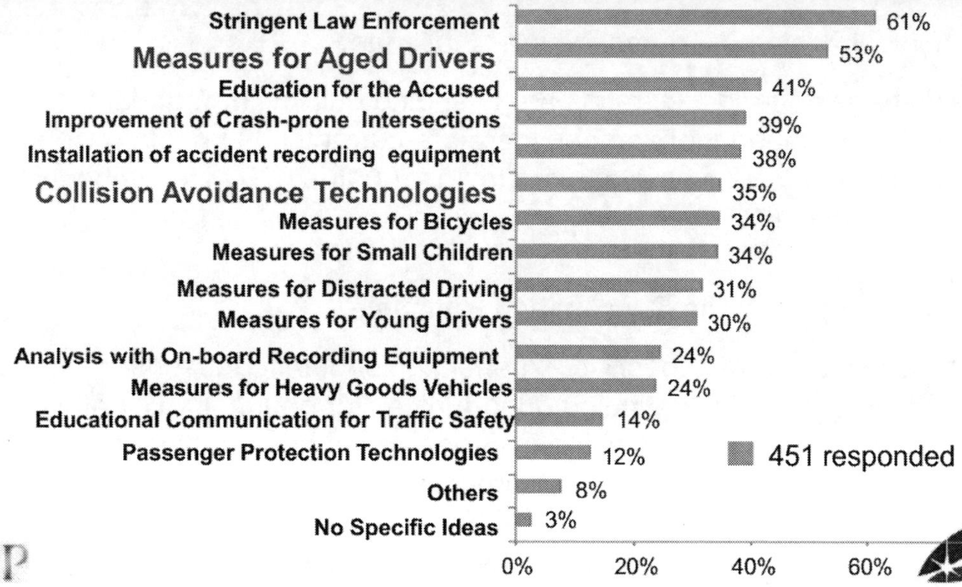

Technologies to evaluate CO_2 reduction

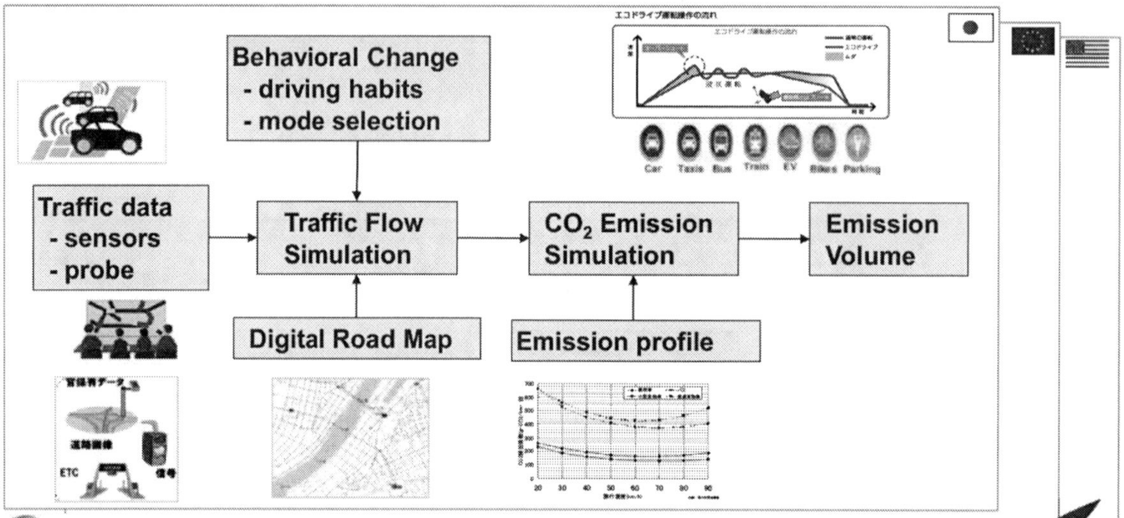

22

WHO: Active Ageing

Global Ageing: A Triumph and a Challenge

Old-age dependency ratios are changing quickly throughout the world. In Japan for example, there are currently 39 people over age 60 for every 100 in the age group 15 – 60. In 2025 this number will increase to 66.

Active Ageing: The Concept and Rationale

If ageing is to be a positive experience, longer life must be accompanied by continuing opportunities for health, participation and security. The World Health Organization has adopted the term “active ageing” to express the process for achieving this vision.

Source: 'Active Ageing, A Policy Framework', World Health Organization, 2002

23

Door to door, staying on the ground level

Mega-cities

- Comprehensive coverage of mass transit systems
- Bus transit systems and pedestrian centered road environment
- Flexible and efficient transit management

Rural towns

- Safe driving assist with automated driving technologies
- Small sized personal mobility vehicles
- On-demand public transportation feasible cost and flexibility

24

ART (Advanced Rapid Transit)

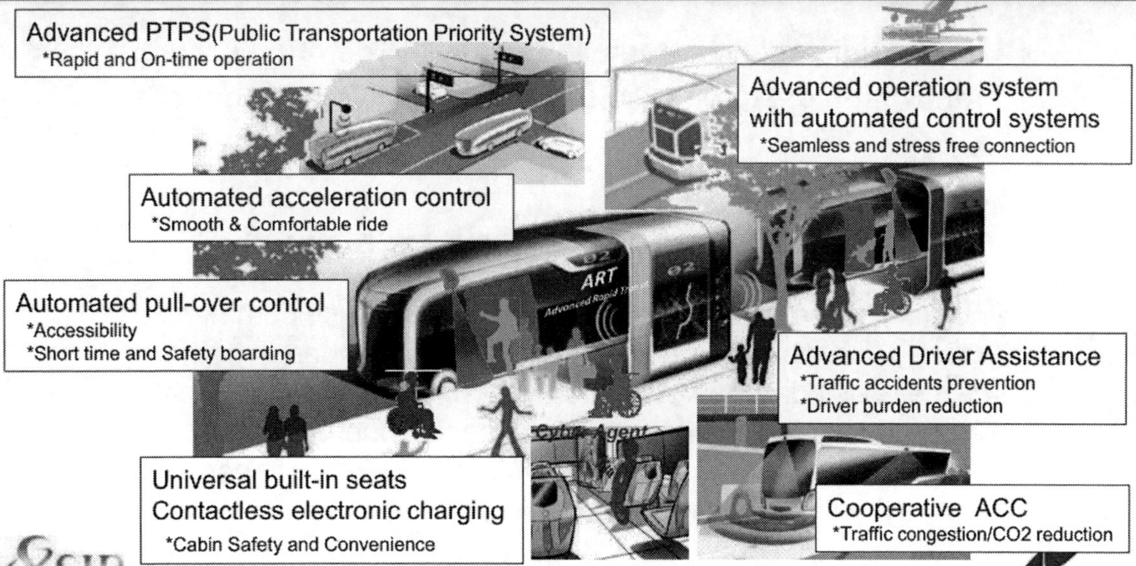

Cross-ministerial Strategic Innovation promotion Program (SIP)

25

Door to door, staying on the ground level

SIP

26

Key Message

Cross-Ministerial Strategic Innovation Promotion program
Innovation of Automated Driving for Universal Services

"SIP- adus"

- Mobility Bringing Everyone a Smile -

Inclusive society, where diverse people in diverse communities actively participate in generating values, will enhance both wellness of individuals and economic development. Automated driving technologies integrated with social innovations should provide everyone with mobility to fully exercise his or her capacity, enabling sustainable development of the society.

SIP

27

1st SIP-adus Workshop on Connected and Automated Driving Systems

SIP

28

2nd SIP-adus Workshop on Connected and Automated Driving Systems

Date: October 27-29, 2015

Venue: Tokyo International Exchange Center

Topics:
1. Dynamic Map
2. Connected Vehicles
3. Human Factors
4. Impact Assessment
5. Next generation transport
6. Security

Program:
- Plenary sessions
- Breakout workshops
- Test rides and demonstrations

SIP

29

European Commission

European Commission

Automated Road Transport
Why cooperate with the European Union?

- EU has strong industrial basis on automotive technologies and systems
- EU has high level of expertise in developing and testing connected and automated vehicle technologies
- Opportunities to tackle challenges of automated road transport together
- Sharing expertise and access to equipment, data and facilities

2

Automated Road Transport - US/EU Cooperation

- EU-US Symposium on Automated Road Transport, 14-15 April 2015, Washington D.C:
  - **50 expert participants (academia, industry, transport operators, service providers, authorities, legal experts, insurers, users)**
  - **Breakout sessions on 3 different use cases**
    - freeway platooning
    - automated city center
    - urban chauffeur
  - **Areas for cooperation**
    - FOT's/Test beds/Demo's (data exchange, best practice guide, model deployment)
    - Standards & certification (minimum performance standards, testing regime)
    - Human factors (acceptance, HMI, VRU, behaviour, safe stops)
    - Legal framework (rules of operation, code of practice, liabilities)
    - Socio-economic impact (evaluation methodology, business models, city planning)
    - Technology (sensors, V2V/I2V, robustness, technical + performance requirements, maps, data)

3

What is Horizon 2020?

- **EU Framework Programme for Research and Innovation**
- **€ 79 billion from 2014 to 2020**
- **Biggest multinational research programme in the world**
- **Covers the full innovation chain**
- **Funds research in all areas of science and innovation**
  - Excellent Science
  - Competitive Industries
  - Tackling global societal challenges

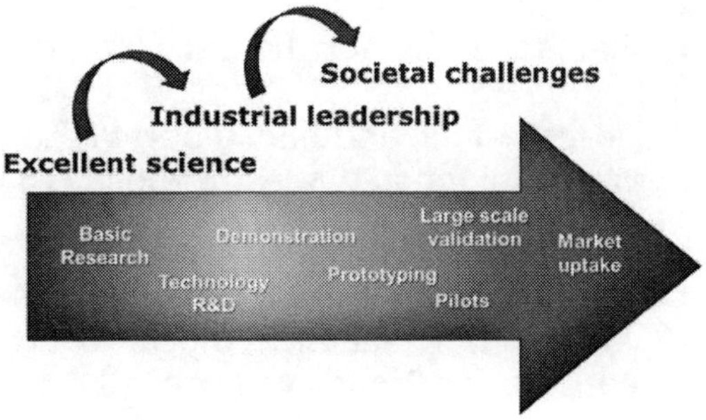

Horizon 2020 – Societal challenges

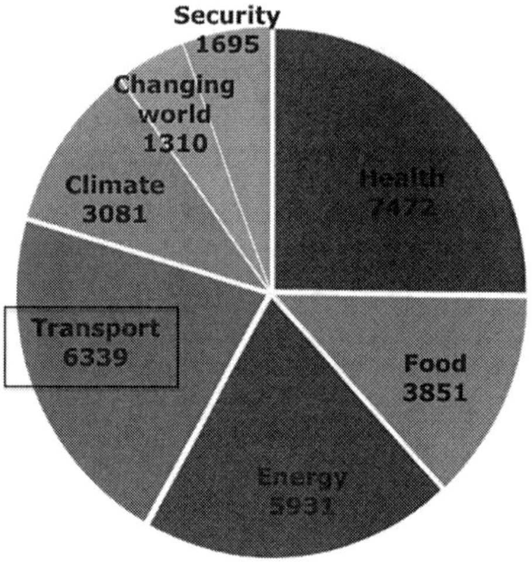

% of H2020

- Health 9,7%
- Food 5%
- Energy 7,7%
- **Transport 8,2%**
- Climate 4%
- Changing world 1,7%
- Security 2,2%

5

Horizon 2020 (2014–2020) - Transport

❑ **Transport is a main priority**
- Budget: 6,339 b€

❑ **4 broad lines of activities**
- Resource efficient transport
- Better mobility, less congestion, more safety and security
- Global leadership for the European transport industry
- Socio-economic and behavioural research and forward looking activities for policy making

❑ **Implemented by**
- Long-term Public-Private Partnerships
- Collaborative Research Projects following Open Calls for Proposals (annually)

6

Horizon 2020 – Transport
CALLS FOR PROPOSALS 2016

- ❑ ***'Mobility for Growth' (MG)***
  1. **Aviation**
  2. **Waterborne**
  3. **Safety**
  4. **Urban**
  5. **Logistics**
  6. **ITS**
  7. **Infrastructure**
  8. **Socio-economics**
- ❑ ***'Automated Road Transport' (ART)***
- ❑ ***'European Green Vehicles Initiative' (EGVIA)***

7

Call: Automated Road Transport – 2016/2017

DRAFT

- ❑ Large Scale Field Operational Tests: Demonstrate the technological readiness, reliability and safety of automated driving functions in large scale pilots at European scale
  - **For passenger cars**
    - integrate and test enabling technologies for passenger cars at automation level 3 (Conditional Automation) and evaluate the benefits in Field Operational Tests (FOTs) at European scale.
    - test automated cars in mixed traffic situations and under different conditions (e.g. traffic intensity, weather, lighting, etc.)
    - analyse interaction between the driver, the cars and the traffic environment and study the behaviour of other traffic participants
  - **Trucks**
    - develop, test and validate platooning concepts, technologies and functionalities
    - demonstrate the robustness of multi-brand platooning using C-ITS communication on a real corridor use case
  - **Urban Transport**
    - demonstrate safety, reliability, and fault tolerance of fully automated urban road transport systems in complex traffic scenarios (with automated and non-automated vehicles, pedestrians, cyclists, powered two-wheelers, etc.)
    - demonstrated systems should be fully integrated into existing public transport systems

8

Call: Automated Road Transport - 2016/2017

DRAFT

- ❑ ICT infrastructure to improve connectivity of automated vehicles
  - ➢ **development, testing and real-life validation of ICT infrastructure architectures, integrating state-of-the-art ICT technologies, systems and functions**
- ❑ User requirements, expectations and concerns
- ❑ Safe human machine interface
- ❑ Safety of automated driving in any kind of complex and mixed traffic situations
- ❑ Innovative modelling, design and engineering of road infrastructure to facilitate the introduction of automated driving systems
- ❑ Specific Coordination and Support Actions

9

Horizon 2020 - Open to the world How can US partners get involved?

- ❑ All Horizon 2020 projects can include international partners
  - ➢ **Applicants from non-EU countries are eligible to take part in Horizon 2020 programmes, even as coordinator**
- ❑ Targeted Opening
  - ➢ **In some topics, identified in the calls, proposals are encouraged to include international partners**
- ❑ Twinning
  - ➢ **In some topics, proposals should foresee twinning with entities participating in projects funded by US DOT**

10

Road vehicle automation - Proposed areas for twinning

- ❑ Knowledge base on all ongoing research and demonstration activities, best practices, automated driving scenarios
- ❑ Strategy for sharing and exploiting collected data in National, European and International FOTs
- ❑ Safety and end-user acceptance aspects of road automation
- ❑ Innovative modelling, design and engineering of road infrastructure
- ❑ Standards & certification (minimum performance standards, testing regime)
- ❑ Full-scale demonstration of urban road transport automation

11

Other proposed areas for twinning

- ❑ **SAFETY**
  - ➢ Protection of all road users in crashes
  - ➢ Innovative design, upgrading and maintenance of Transport infrastructure to increase the transport system safety at modal and intermodal level
  - ➢ Behavioural aspects for safer transport
- ❑ **ITS**
  - ➢ Large-scale demonstration(s) of cooperative ITS
  - ➢ Roadmap, new business models, awareness raising, support and incentives for the roll-out of ITS
- ❑ **INFRASTRUCTURE**
  - ➢ Resilience to extreme (natural and man-made) events
  - ➢ Optimisation of transport infrastructure multi-modal corridors and terminals

12

European Commission

US/EU Suggested Twinning Process

- Agree on common areas for twinning/cooperation
- European Commission publishes call for proposal encouraging twinning / US Funding Agencies may match it with a similar request for research
- US and EU Funding Agencies identify projects for twinning and establish first contacts
- Projects on both sides agree on twinning activities and respective efforts

Please note:

- Collaboration could consist of exchanges of information, data, visits, methodologies, researchers, results, joint workshops, publications; etc.
- No merging of services, resources (including financial) is foreseen > no payments across the Atlantic;

13

European Commission

TRA 2016: 18-21 April 2016, Warsaw

6th European Conference on Transport Research (TRA)
Warsaw (Poland), 18-21 April 2016

Moving forward: Innovative Solutions for Tomorrow's Mobility
www.traconference.eu

European Commission

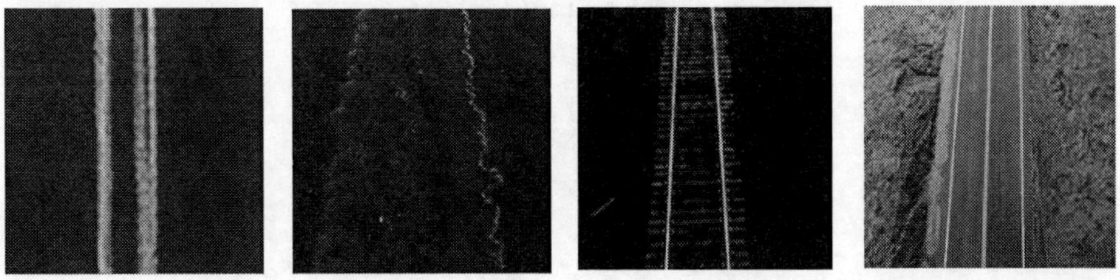

Thank you for your attention!

Ludger ROGGE
European Commission, Brussels/Belgium
DG Research & Innovation
E-MAIL: ludger.rogge@ec.europa.eu

Find out more:

www.ec.europa.eu/research/horizon2020
www.ec.europa.eu/research/participants/portal/page/home

15

U.S. Department of Transportation
Office of the Assistant Secretary for Research and Technology

U.S. DOT Automation Program

Kevin Dopart
ITS Joint Program Office, OST-R
U.S. Department of Transportation
July 23, 2015

U.S. Department of Transportation
Office of the Assistant Secretary for Research and Technology

Automation within DOT

CONTEXT

Context: ITS Joint Program Office

The ITS JPO has Department-wide authority in coordinating the ITS program and initiatives among the following DOT Offices:

- Federal Highway Administration (FHWA)
- Federal Motor Carrier Safety Administration (FMCSA)
- Federal Transit Administration (FTA)
- Federal Railroad Administration (FRA)
- National Highway Traffic Safety Administration (NHTSA)
- Maritime Administration (MARAD).

U.S. Department of Transportation
ITS Joint Program Office 3

Context: ITS Strategic Plan Framework

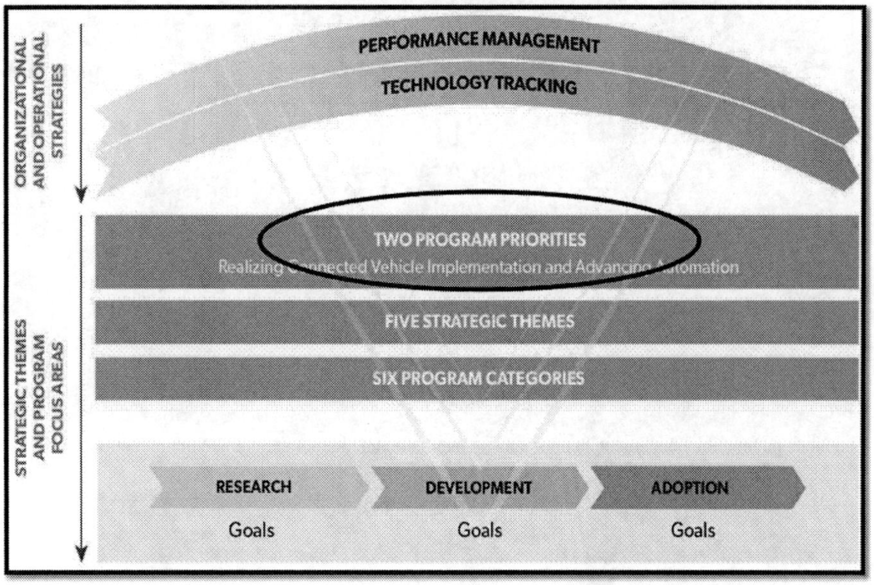

U.S. Department of Transportation
ITS Joint Program Office 4

Trends: How We Move ...

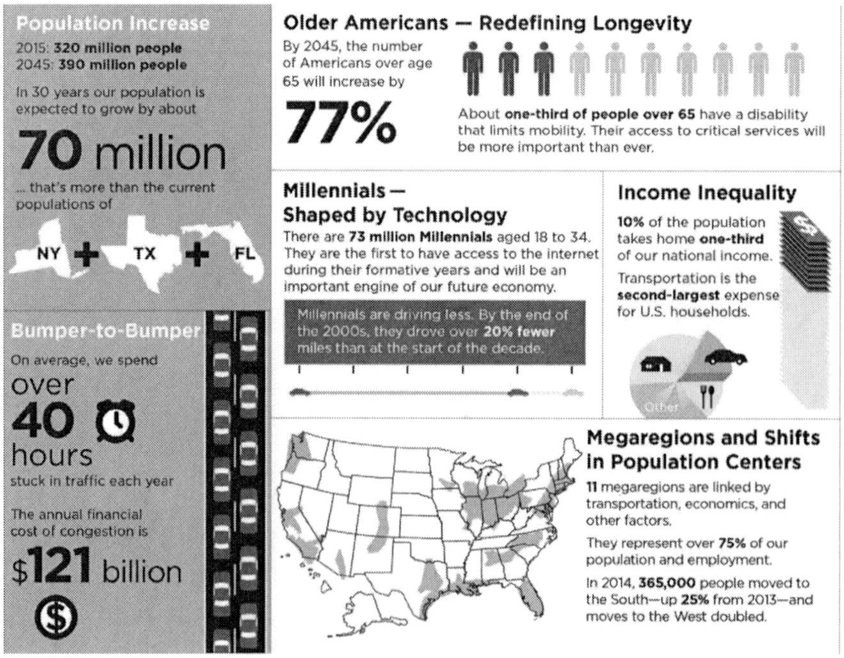

6

Context: Automation in Beyond Traffic

Automation will have a potentially transformative impact across all transportation modes, increasing productivity, improving safety, and enhancing the capacity of existing infrastructure. It may also have a profound impact on the transportation workforce, changing the skills required to manage, operate, and maintain transportation vehicles and systems.

U.S. Department of Transportation
ITS Joint Program Office

7

U.S. Department of Transportation
Office of the Assistant Secretary for Research and Technology

Foundation

CONNECTED AUTOMATION

Automation Can Be a Tool for Solving Transportation Problems

- Improving safety
  - Reduce and mitigate crashes
- Increasing mobility and accessibility
  - Expand capacity of roadway infrastructure
  - Enhance traffic flow dynamics
  - More personal mobility options for disabled and aging population
- Reducing energy use and emissions
  - Aerodynamic "drafting"
  - Improve traffic flow dynamics

...but connectivity is critical to achieving the greatest benefits

U.S. Department of Transportation
ITS Joint Program Office 9

Connected Automation for Greatest Benefits

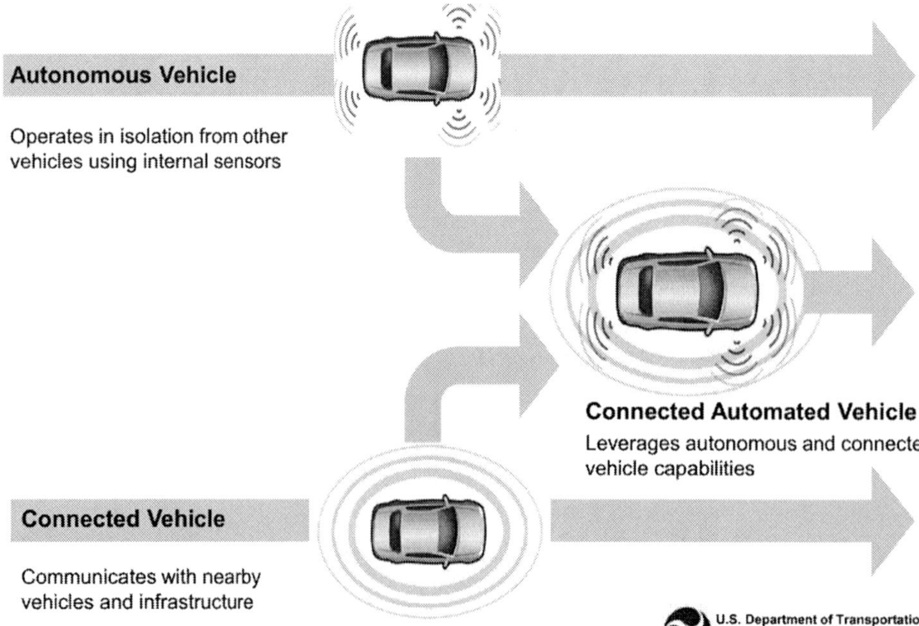

U.S. Department of Transportation
ITS Joint Program Office 10

Connected Vehicle Milestones

- August 2014: NHTSA ANPRM on vehicle-to-vehicle communications
- May 2015: Secretary Foxx V2V announces V2V rulemaking acceleration
- Summer 2015: FHWA V2I guidance document
- Fall 2015: First wave of CV Pilots to begin
- End of 2015: V2V NPRM interagency review
- New cars with connected vehicle technology are expected to be available by 2017.

U.S. Department of Transportation
ITS Joint Program Office 11

Connected Vehicle Pilot Deployment Program

CV Pilot Program Goals

Proposed Program Schedule

- September 2015 - Wave 1 Pilot Deployments Award(s)
- Early 2017 - Solicitation for Wave 2 Pilot Deployment Concepts
- September 2017 - Wave 2 Pilot Deployments Award(s)
- September 2020 - Pilot Deployments Complete

Resources

- ITS JPO Website: http://www.its.dot.gov/
- CV Pilots Program Website: http://www.its.dot.gov/pilots

U.S. Department of Transportation
ITS Joint Program Office 12

2015 FHWA Guidance Will Help Communities Prepare for Connected Vehicles

- The FHWA is developing policy positions, guidance, guidelines, whitepapers, and practitioner tools to promote the smooth deployment of V2I technology by transportation system owners/ operators.
- The FHWA will issue initial guidance in late 2015. This initial guidance is intended to assist in planning for future investments and deployment of V2I systems.
- The guidance does not impose any new requirements on local governments.
- This work will be harmonized with related efforts by other USDOT modal agencies.
- Subsequent guidance updates will also incorporate ITS research findings.

U.S. Department of Transportation
ITS Joint Program Office 13

5.9 GHz Spectrum Sharing

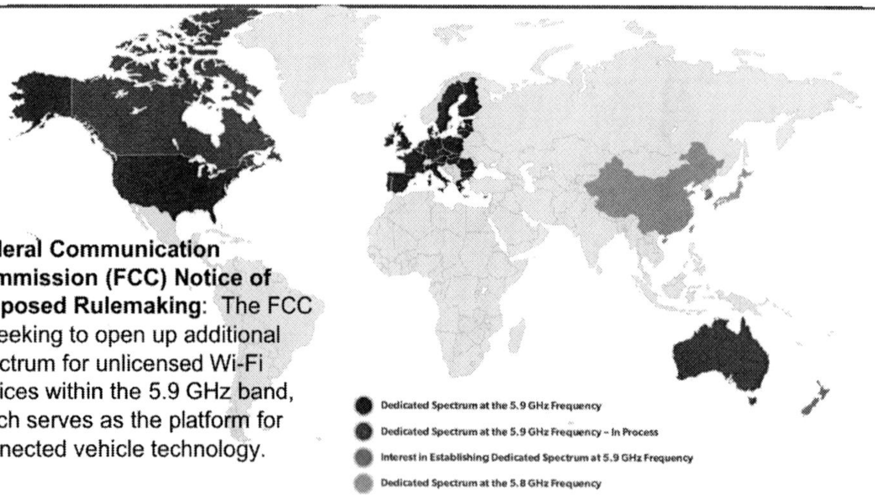

- **Federal Communication Commission (FCC) Notice of Proposed Rulemaking**: The FCC is seeking to open up additional spectrum for unlicensed Wi-Fi devices within the 5.9 GHz band, which serves as the platform for connected vehicle technology.

- **5.9 GHz Spectrum:** The connected vehicle environment that is being researched is based on reliable access to the 5.9 GHz wireless spectrum.
- **Spectrum Sharing:** Any changes to the 5.9 GHz spectrum may jeopardize crash avoidance capabilities.

U.S. Department of Transportation
ITS Joint Program Office 14

U.S. Department of Transportation
Office of the Assistant Secretary for Research and Technology

Overview

U.S. DOT AUTOMATION PROGRAM

U.S. DOT Automation Program

Goal: Enable safe, efficient, and equitable integration of automation into the transportation system

Area	Example Applications	Research Emphasis
Connected Driving Assistance *Level 1-2*	Platooning, merge/weave assist, speed harmonization, and eco-approach and departure	**Benefits** (safety, mobility, sustainability) and **Application Development**
Conditional Automation *Level 2-3*	Highway autopilot, traffic jam assist, etc.	**Safety Assurance** (human factors, control system reliability, testing procedures, and cybersecurity)
Limited Driverless Vehicle Operations *Level 4*	Low-speed automated shuttles, first-last mile transportation	**Feasibility** (concept development, testing, evaluation)

U.S. Department of Transportation
ITS Joint Program Office

16

Research Tracks

Enabling Technologies
- Digital Infrastructure
- Communications
- Technology Research

Safety Assurance
- Electronic Control Systems
- Functional Safety and Electronics Reliability
- Cybersecurity
- Human Factors

Transportation System Performance
- CACC, Speed Harmonization, and Platooning
- Lateral Control
- First/Last Mile and Transit Operations

Testing and Evaluation
- Interoperability
- Testing Methods
- Benefits Assessment

Policy and Planning
- Standards
- Federal Policy Analysis
- Stakeholder Engagement
- Transportation Planning

U.S. Department of Transportation
ITS Joint Program Office 17

Stakeholder Engagement

- Automated Vehicle Symposium
  - Proud supporter (2012-present)
  - USDOT Listening Session - Thursday 1:30-3:00 PM
- Roundtable Discussions
  - Early Adopter States
- Webinars
  - ITS JPO and ITS America Webinar: Fundamental Issues for Road Transport Automation
  - ITS PCB Talking Transportation and Technology (T3) Webinar Series
  - https://www.pcb.its.dot.gov/t3_webinars.aspx
- NCHRP 20-102
- Coordination with associations (e.g. AASHTO, AAMVA)
- Inviting external research briefings into DOT
- ....and more to come!

AUTOMATED VEHICLES SYMPOSIUM 2015
DRIVERS. VEHICLES. INFRASTRUCTURE.

ITS AMERICA

TRB
TRANSPORTATION RESEARCH BOARD
OF THE NATIONAL ACADEMIES

U.S. Department of Transportation
ITS Joint Program Office 18

International Coordination

- Trilateral Working Group on Automation in Road Transportation
  - European Union
  - Japan
  - United States
- Complementary EU-US Research Programming on selected issues of shared interest.

EU★US★JAPAN
ITS COOPERATION

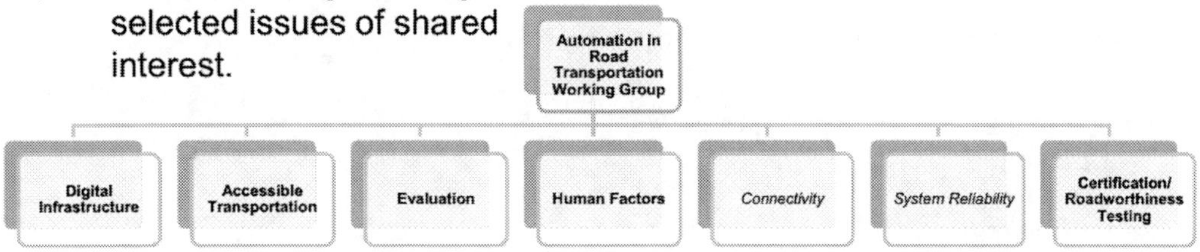

U.S. Department of Transportation
Office of the Assistant Secretary for Research and Technology

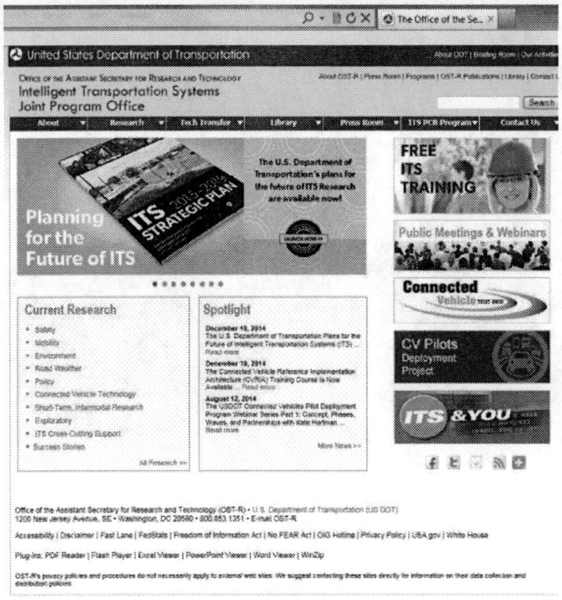

www.its.dot.gov

Kevin Dopart
US DOT / ITS JPO
Kevin.Dopart@dot.gov

Sustainable TRANSPORTATION

U.S. DEPARTMENT OF ENERGY | Energy Efficiency & Renewable Energy

Transportation as a System

Getting SMARTer on Energy & Mobility

AVS 2015 - July 23, 2015

Reuben Sarkar
Deputy Assistant Secretary - Transportation
Energy Efficiency and Renewable Energy
U.S. Department of Energy

Transportation Energy & The Environment

- 70% of all U.S. petroleum use
- 27% of U.S. greenhouse gas emissions (GHG)
- 1.8 Gigatons of CO_2-equivalent

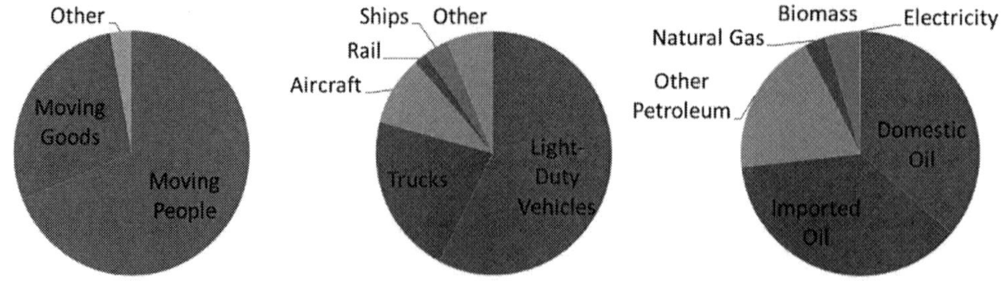

26.8 Quadrillion Btu of transportation energy use

(note: does not include approximately 2.2 quadrillion Btu for offroad equipment)

Transportation is still 93% dependent on petroleum

Major Administration Energy Goals

Reduce GHG emissions by 17% by 2020, 26-28% by 2025 and 83% by 2050 from 2005 baseline Climate Action Plan

Reduce net oil imports by half by 2020 from a 2008 baseline Blueprint Secure

Double energy productivity by 2030 Department of Energy

By 2035, generate 80% of electricity from a diverse set of clean energy resources Blueprint Secure Energy Future

Reduce CO_2 emissions by **3 billion metric tons** cumulatively by 2030 through efficiency standards set between 2009 and 2016 CAP Progress Report

3

U.S. DEPARTMENT OF ENERGY | Energy Efficiency & Renewable Energy

Sustainable TRANSPORTATION Technology Offices

$793M FY16 ($602M FY15)

BIOENERGY

Drop-in Biofuels
Bioproducts & Biopower
Feedstocks & Logistics
Conversion
Demonstration
Market Transformation
Sustainability

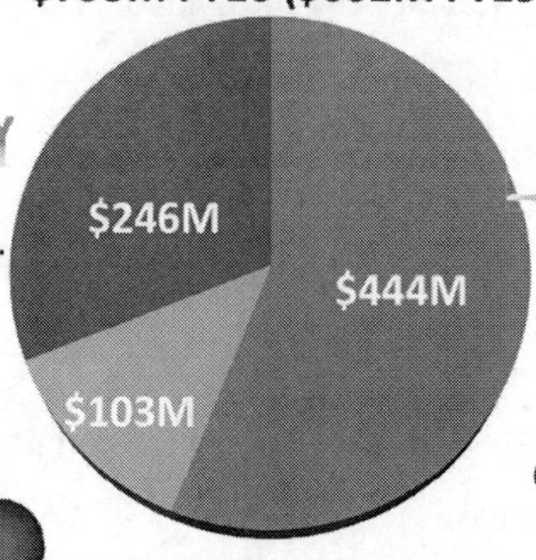

VEHICLES

Batteries
Electric Drives
Advanced Combustion
Vehicle Systems
Material Technology
Fuels & Lubricants
Outreach & Deployment
Analysis

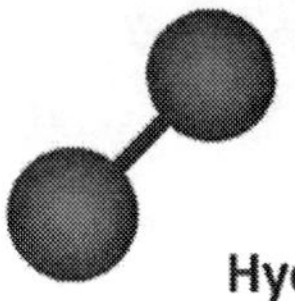

HYDROGEN & FUEL CELL

Fuel Cell Systems
Hydrogen Production & Delivery
Safety Codes & Standards
Technology Validation
Market Transformation

4

U.S. DEPARTMENT OF ENERGY | Energy Efficiency & Renewable Energy

Transportation as a System

Today:
- Vehicle level focus
- Independent
- Unconnected
- Subject to behaviors & decisions

Tomorrow:
- System level focus
- Connected
- Automated
- In concert
- Across modes
- Managed behaviors & decisions

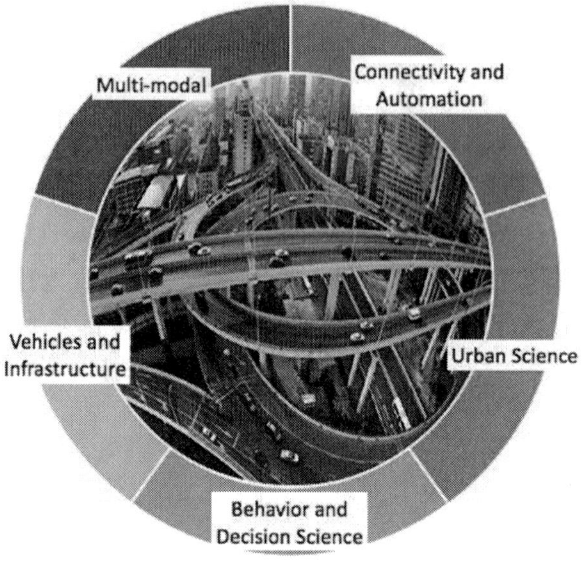

Exploring the untapped transportation system level efficiencies

Connectivity, Automation and Energy

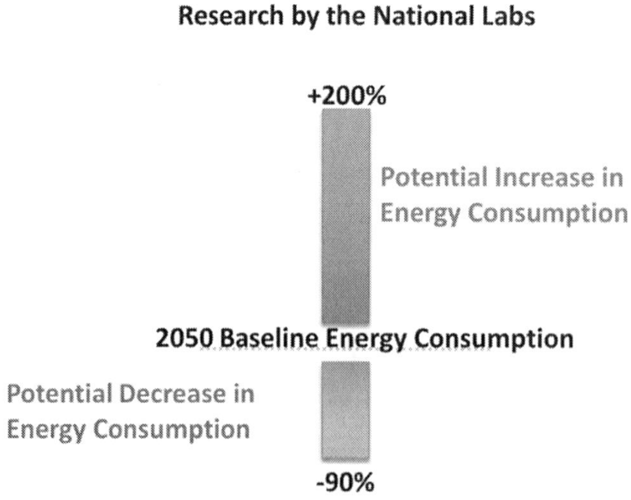

Vast range of energy implications ... more research required

Connected & Automated Vehicles (CAVs)

Expected Outcomes:

- Quantify the energy impact of CAVs
  - Multi-scale
  - Multiple scenarios
  - Different technologies
- Identify CAV-enabled opportunities
  - Vehicle electrification
  - Lightweighting
  - Powertrain optimization
  - Vehicle utilization
  - Reduced VMT's
- Inform policy/research on CAV's
  - Maximize sustainability impacts

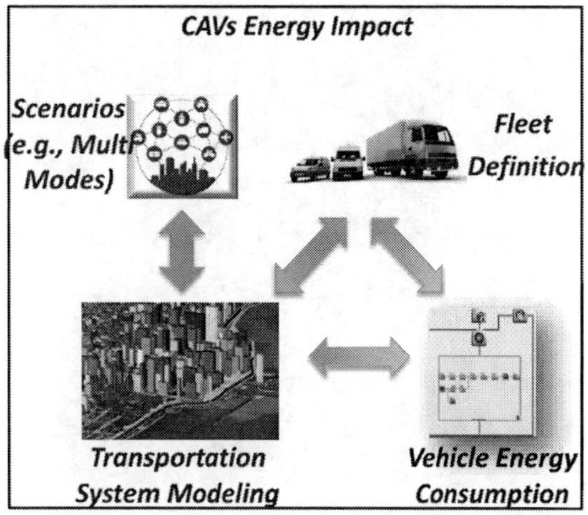

Designing for the nexus of safety, energy, and mobility

Vehicles and Infrastructure

Expected Outcomes:

- Integrated modeling of vehicle and fuel technologies with consumer preferences
- Best leverage public and private resources for EV/AFV fueling infrastructure

Mapping EV Technology with Travel Patterns Reduced EVSE Locations from 18,000+ to 281 in Seattle

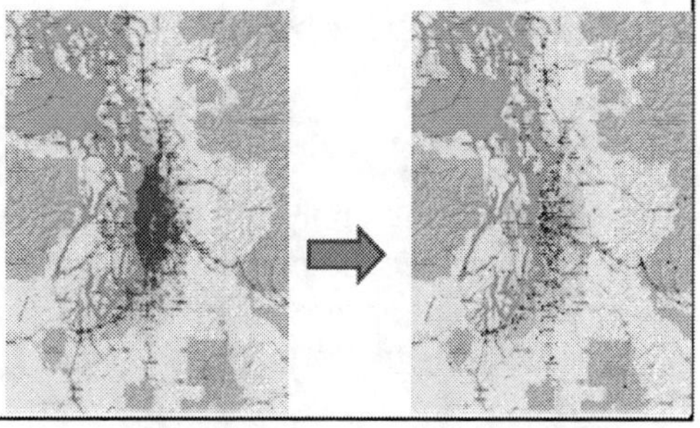

Wood, E; Burton, E; and Neubauer, J. (2015). *"Measuring the Benefits of Public Chargers and Improving Infrastructure Deployments Using Advanced Simulation Tools".*

Informed infrastructure investments that drive consumer adoption

Multi-Modal

Expected Outcomes:

- Quantify potential energy savings and GHG reductions in urban areas
  - Diminished modal barriers
  - Passenger and freight
- Counteract projected growth of freight energy consumption (through 2040)
  - Leveraging of disparate modal energy intensities
  - Streamline transfers, shift to new modes, etc.

(Photo credits: AP/Elaine Thompson, RailNewYork.com, CNN.com, Jeffrey Beall, Wikimedia Commons)

Energy-efficient, seamless multi-modal transport of people and goods

Urban Sciences

Expected Outcomes:

- New city-scale computational models
  - Calibrated and validated by large transportation data sets
  - Can inform local decisions
- Frameworks and analytical tools
  - Build and run composite models of urban components related to sustainable transportation

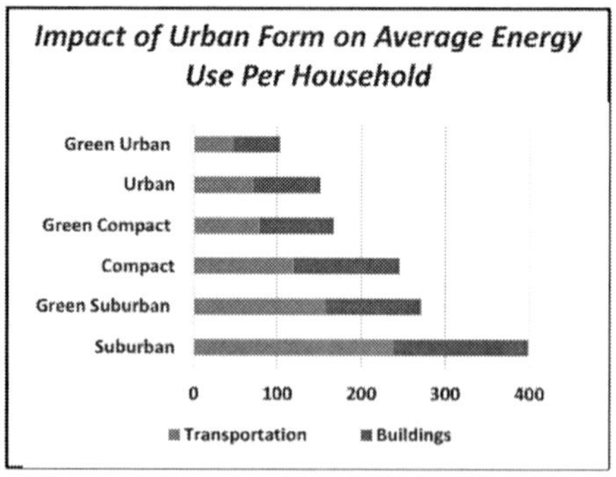

Calthorpe, P. Urbanism in the Age of Climate change. November 2010.

Aligning urban form with sustainable transportation

Behavioral and Decision Science

Expected Outcomes:

- Enhanced vehicle adoption and choice models
    - Inform holistic policy decisions, vehicle R&D, and infrastructure investments
    - Accelerate PEV adoption
- Understanding of individual and market behavior
    - Future technologies, policies, and transportation systems

Transportation System Decision Points

Technology & policies that anticipate how decisions are really made

U.S. DEPARTMENT OF ENERGY | Energy Efficiency & Renewable Energy

Simulation & Modeling Tools

Independent Vehicle

Small Network of CAVs and ITS
"Virtual Testing Ground"

Entire Urban Area
Transportation Systems Energy Model

Courtesy of Argonne National Labs

Integrated multi-scale tools for local, regional to national scale systems

U.S. DEPARTMENT OF ENERGY | Energy Efficiency & Renewable Energy

Semi-Automated Platooning

- **NREL, Peloton*, Intertek, PACCAR**
- American-style line haul sleeper cabs
  - Modern aerodynamics
  - EPA SmartWay tractors; trailers with side skirts
- Enabling technologies for platooning*
  - Forward object detection (radar, laser, stereo cameras, etc.)
  - Dedication short-range communication (DSRC)
  - Vehicle-to-vehicle communications (V2V) and driver displays
  - Vehicle braking and torque control interface
- SAE J1321 Type II Fuel Consumption Test Procedure
- 20 to 75 ft vehicle gaps, Up to 70 mph, 65,000 lbs

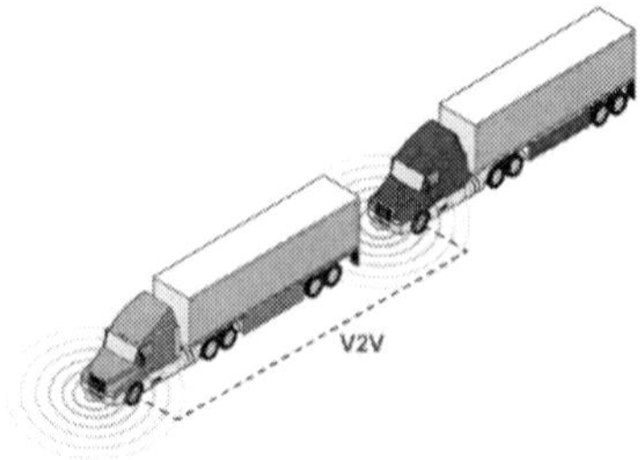

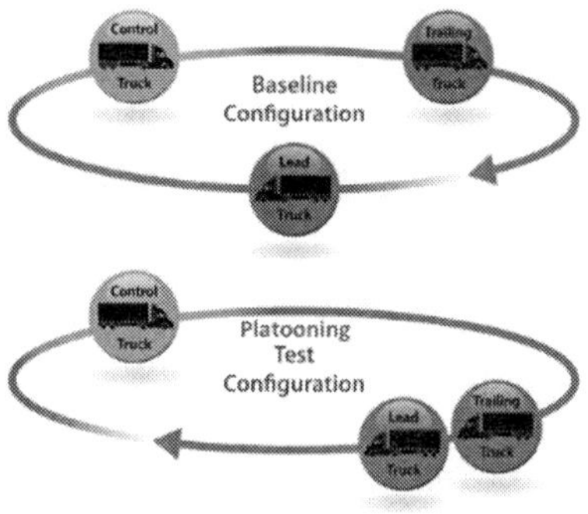

13 U.S. DEPARTMENT OF ENERGY | Energy Efficiency & Renewable Energy

Fuel Savings Potential

- Up to **9.7%** fuel savings for the trailing truck
- Up to **5.3%** fuels savings for the lead truck

Team fuel savings ranging from 3.7% – 6.4%

14 U.S. DEPARTMENT OF ENERGY | Energy Efficiency & Renewable Energy

Coming Soon...

Systems and
Modeling for
Accelerated
Research in
Transportation

- National Labs
- Federal Agencies
- State & Local Governments
- Universities
- Industry

Focus on Energy & Mobility

... DOE SMART Mobility Consortia

15

U.S. DEPARTMENT OF ENERGY | Energy Efficiency & Renewable Energy

System of Systems

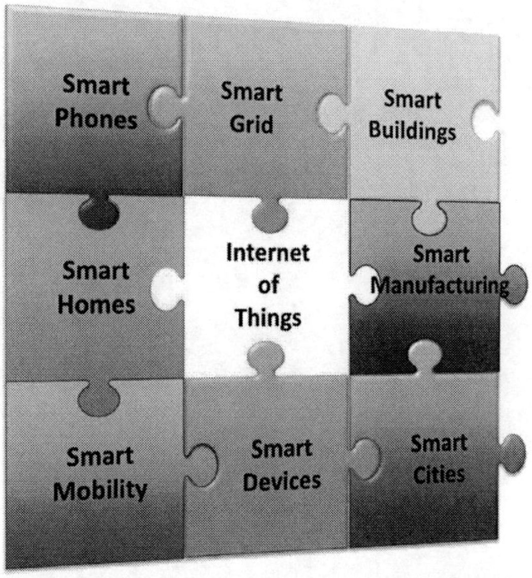

We are thinking about the Federal Role in the Internet of Things ...

16

U.S. DEPARTMENT OF ENERGY | Energy Efficiency & Renewable Energy

Questions

Reuben.Sarkar@ee.doe.gov

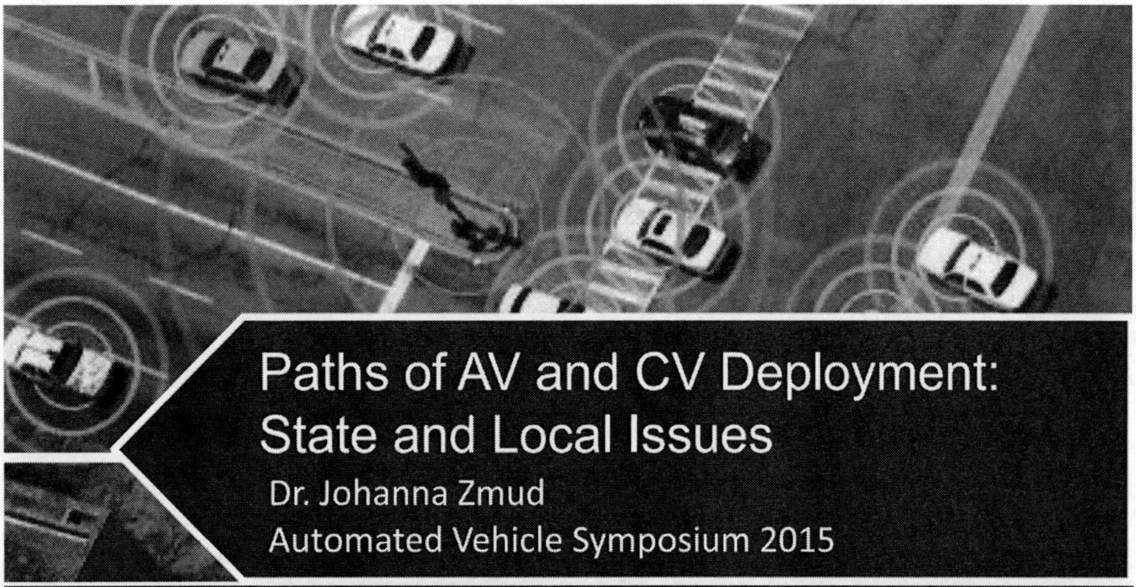

Paths of AV and CV Deployment: State and Local Issues

Dr. Johanna Zmud

Automated Vehicle Symposium 2015

Texas A&M Transportation Institute

1

Scenario-Based Roadmapping

- Scenarios of paths of AV CV deployment
  - Literature review
  - Expert workshops
- Implications and impacts
  - Interviews with state and local transportation staff
- Strategic roadmap
  - TTI internal workshop

Texas A&M Transportation Institute

State and Local Perspectives on AV

- Driven by OEM/private sector
- Disruptive: What's expected of us?
- Uncertainty: mixed traffic
- Issues: construction zones, snow/ice, signage/ striping, traffic signals

Texas A&M Transportation Institute

State and Local Perspectives on CV

- OEMs/private sector will implement V2V
- Unsure about V2I - **$**
  - Implementation
  - Ongoing tech support
  - Databases and detailed mapping
- Data from CV applications
  - Excitement
  - Ownership questions

Texas A&M Transportation Institute

Scenario Approach

- Distinct narratives about future
- Represent extremes
- Plausible

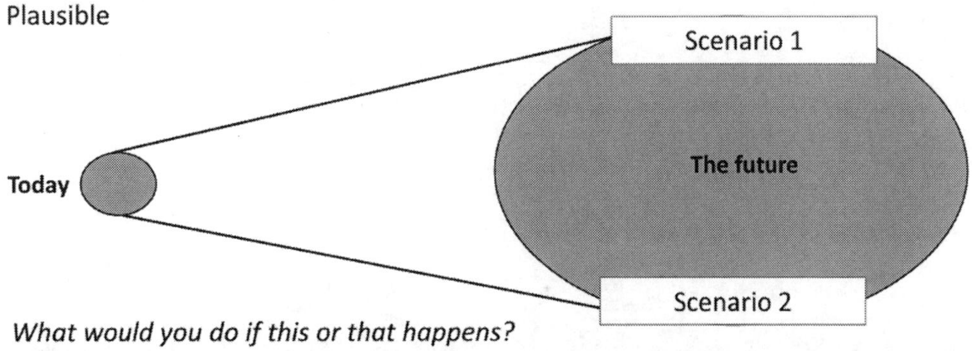

- *What would you do if this or that happens?*

Texas A&M Transportation Institute

Systems Thinking: Influencing Areas

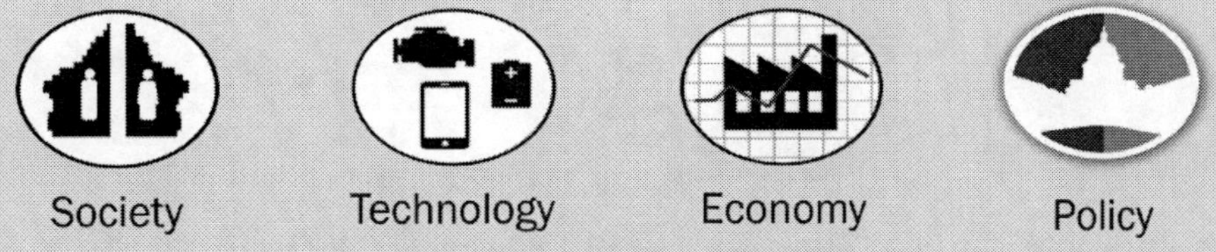

Evolutionary Scenario

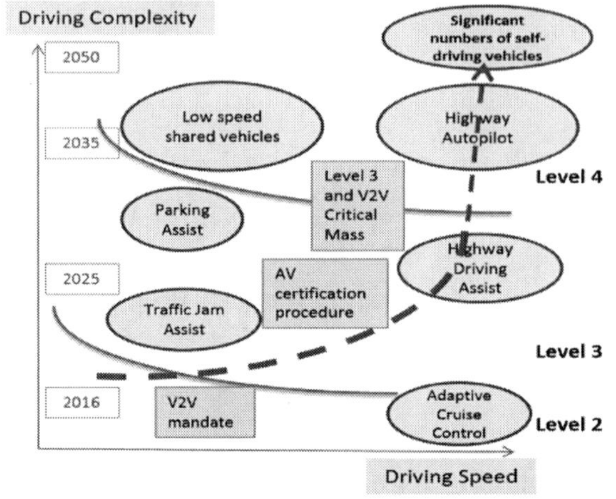

Texas A&M Transportation Institute

Revolutionary Scenario

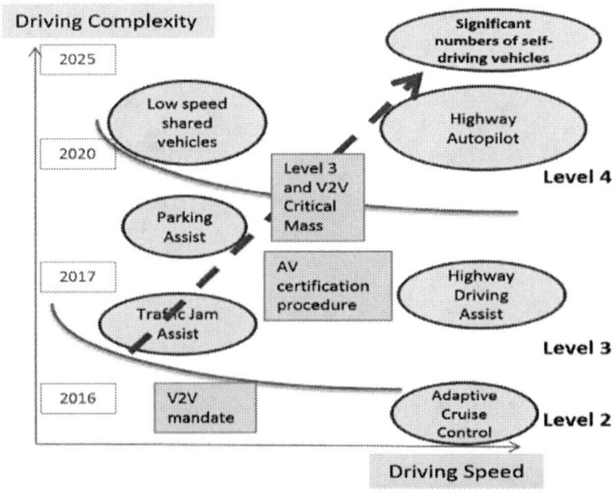

Texas A&M Transportation Institute

Reactions to Scenarios

Scenario	Likelihood	Preference
Evolutionary	Regulatory change and fleet turnover slow even though technology changes fast 51%	Easier to adapt, less stressful More time to evolve the enabling infrastructure 38%
Revolutionary	AV – OEMs pushing, Consumers buying Quick use cases: Trucking, shared ride, package delivery 49%	Private sector push brings capital resources to make it happen Benefits evident and should be realized as quickly as safely to do so 62%

Texas A&M Transportation Institute

Potential Changes for Organization

- Mission
  - No change
- Responsibilities

  ⬇ construction, safety, human services transportation, traditional ITS, parking

  ⬆ maintenance, operations, "big data" management and analysis
- Structure
  - Larger operations group
  - AV/CV section, with representation from existing groups

Texas A&M Transportation Institute

Policy or Planning Actions to Prepare

- Positioning for economic development potential
- Work with DMV on licensing implications
- Incorporating AV into planning efforts
- Modifying roads/ enabling infrastructure
- How/ where of dedicated lanes for transition (mixed fleets)
- Be intentional about early deployments
- None – be flexible as/if technology changes

Texas A&M Transportation Institute

Thank you!

Research Team:
Johanna Zmud j-zmud@tti.tamu.edu
Melissa Tooley
Trey Baker
Jason Wagner

Texas A&M Transportation Institute

FLORIDA AUTOMATED VEHICLES

Creating the Framework for Implementation

State and City Level Issues

Secretary Paul Steinman (District 7)

Centennial FDOT 1915 • 2015

Implementation Challenges of Connected Automation

Making Business Decisions
- Requires new benefit/cost analysis to support deployment decisions
- Needs systematic & strategic approach

New Investments Needed
- Funding sources
- Infrastructure requirements
- Staffing needs

Data Issues
- Ownership
- Privacy/security
- Access & support

Interoperability
- Local, regional, national
- Multiple platforms & protocols

Public Sector Perspective

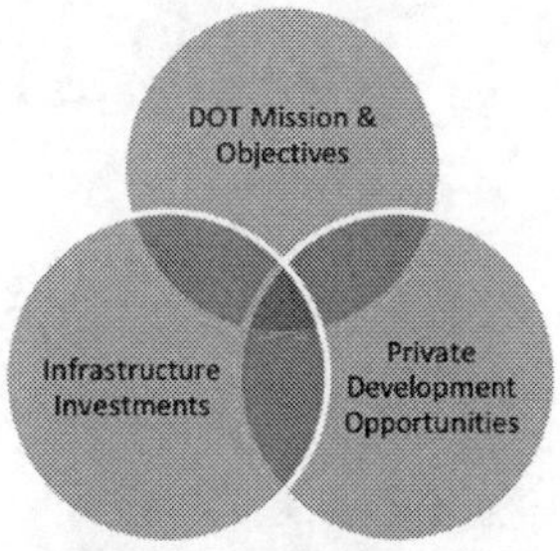

Land Use Impacts

Wide scale adoption of AV/CV technologies will alter transportation usage, site development, and ultimately land use patterns.

- Drop off zones
- Vehicle lane design standards
- Parking location
- Enhanced multi-modal options
- Ride sharing
- Commuting patterns

AASHTO = Central Role in Coordinating State Efforts

Interoperability
Standard protocols for software/ hardware - requirements for seamless communication between states and regions

Reciprocity
Consistent state statutes and regulations to allow for AV/CV technologies to operate coast-to-coast

Testing and Pilot Agreements
If state laws are not consistent, states should be agreeable to support inter-state testing and pilot projects

Avoid Piecemeal Regulations
AASHTO should serve leadership role in coordinating efforts amongst the states to provide USDOT with best practices

US House Committee on Transportation and Infrastructure has expressed concern

USDOT & State DOT/DMV Roundtable

Automated Vehicles and the States (May 19, 2015)

Attendance (2 reps from each state):

1) California
2) Florida
3) Michigan
4) Nevada
5) Pennsylvania
6) District of Columbia
7) USDOT modal agencies
8) Key associations
   (AASHTO, NHTSA, AAMVA, GHSA)

Agenda:

- Introductions and State Updates
- Priorities and Challenges
- Policy/Technical/Regulatory
- From a DOT and DMV perspective
- Updates from NHTSA
- USDOT Automation Program Overview
- Next Steps and Recommendations
  - Where is Federal guidance, research, and leadership most helpful?
  - How should states and the USDOT engage going forward?

FDOT AV/CV/ITS Steering Committee

Earlier this year, Secretary Jim Boxold established a Steering Committee comprised of key FDOT stakeholders, led by Assistant Secretary Rich Biter, to coordinate and provide leadership direction over FDOT's AV/CV/ITS initiatives.

Florida Automated Vehicles

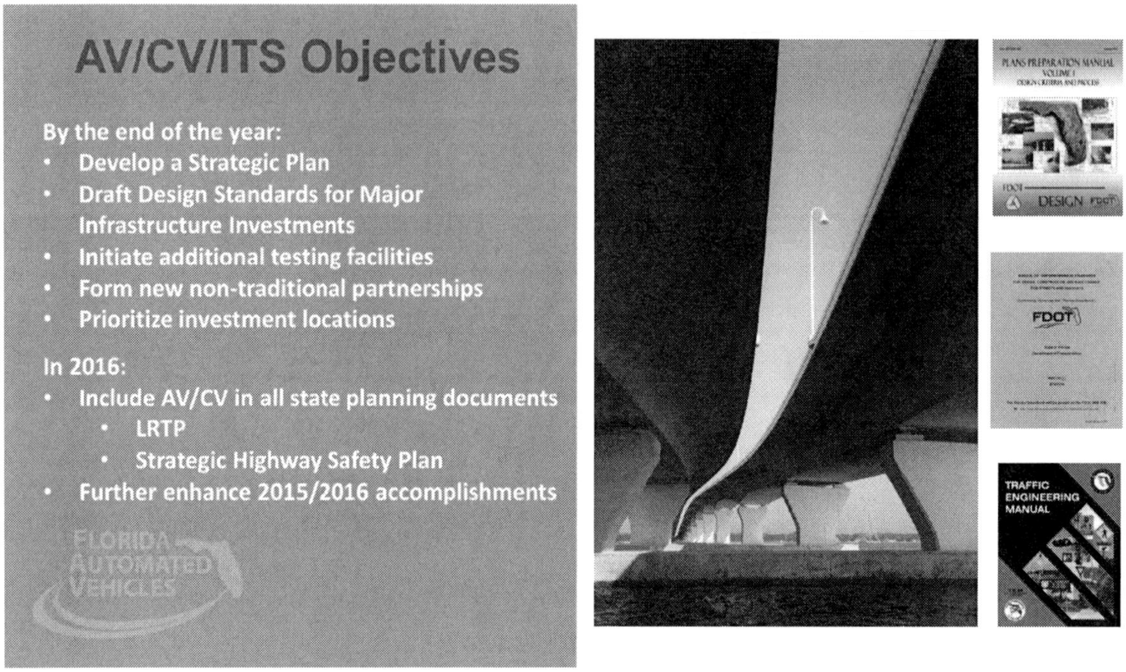
AV/CV/ITS Objectives
By the end of the year:
• Develop a Strategic Plan
• Draft Design Standards for Major Infrastructure Investments
• Initiate additional testing facilities
• Form new non-traditional partnerships
• Prioritize investment locations
In 2016:
• Include AV/CV in all state planning documents
• LRTP
• Strategic Highway Safety Plan
• Further enhance 2015/2016 accomplishments
FLORIDA AUTOMATED VEHICLES
PLANS PREPARATION MANUAL
DESIGN
FDOT
TRAFFIC ENGINEERING MANUAL

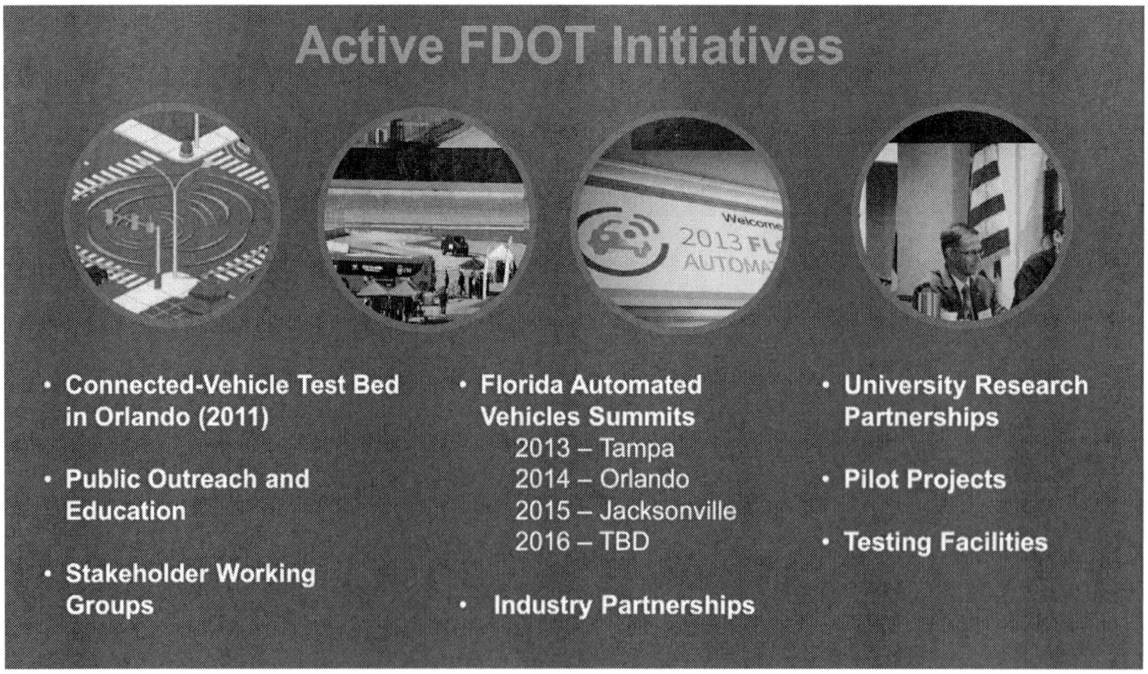
Active FDOT Initiatives
• Connected-Vehicle Test Bed in Orlando (2011)
• Public Outreach and Education
• Stakeholder Working Groups
• Florida Automated Vehicles Summits
2013 – Tampa
2014 – Orlando
2015 – Jacksonville
2016 – TBD
• Industry Partnerships
• University Research Partnerships
• Pilot Projects
• Testing Facilities

Stakeholder Working Groups

1) Policies & Legal Issues
   - Investment funding
2) Infrastructure/Technology
   - New roadway infrastructure
   - Engineering & design standard updates
3) Modal Applications
   - Transit
   - Freight
   - Infrastructure Inspections

Pilot Projects

- Focused on reducing the frequency and severity of crashes
  80% of all avoidable collisions could be prevented using CV technologies
- Commercial vehicle applications
  Improved intermodal connectivity
  Reduced bottlenecks at ports
  Increased safety at intersections
- Connected-Automated (level 4) trucks for drayage operations
  Severe shortage of drivers
  Repetitive low speed routes near ports

Freight Applications for CV Pilot Project

Assessing Connected Vehicle Technologies for Miami's Perishable Freight Industry

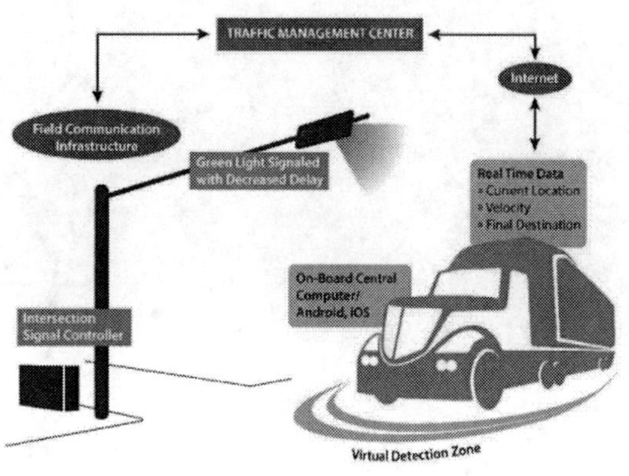

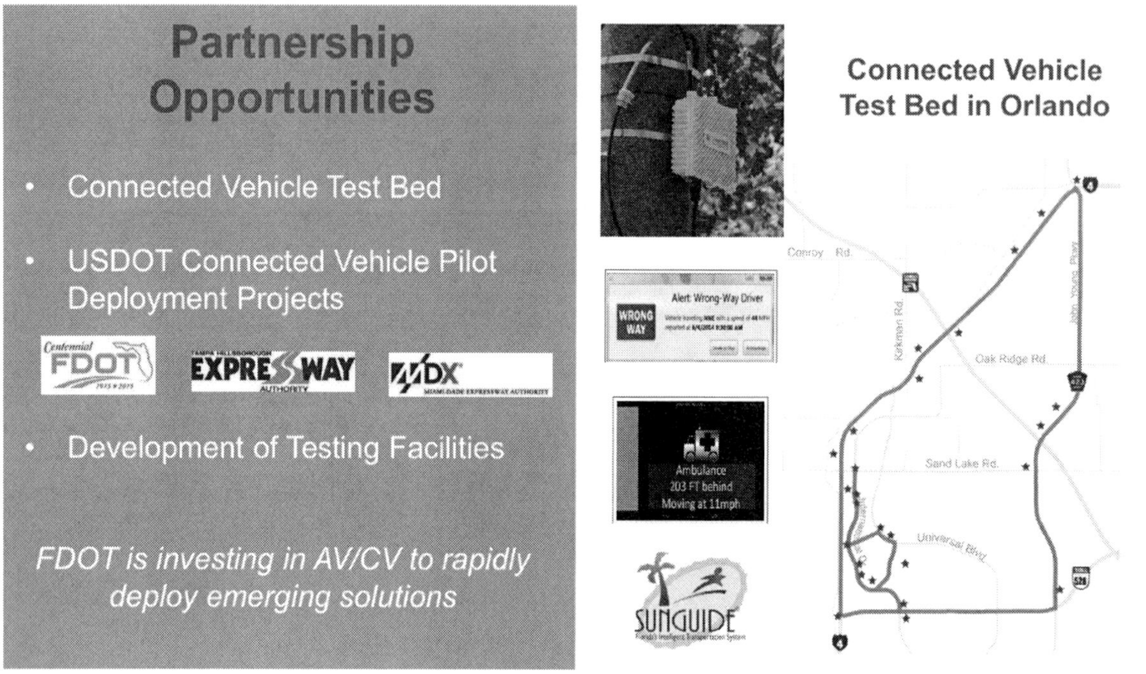
Partnership Opportunities
• Connected Vehicle Test Bed
• USDOT Connected Vehicle Pilot Deployment Projects
FDOT
EXPRESSWAY
MDX
• Development of Testing Facilities
FDOT is investing in AV/CV to rapidly deploy emerging solutions
Alert: Wrong-Way Driver
WRONG WAY
Ambulance
203 FT behind
Moving at 11mph
SUNGUIDE
Connected Vehicle Test Bed in Orlando
Conroy Rd.
Kirkman Rd.
John Young Pkwy
Oak Ridge Rd.
Sand Lake Rd.
Universal Blvd

Thank you
FLORIDA DEPARTMENT OF TRANSPORTATION PRESENTS
FLORIDA AUTOMATED VEHICLES
SUMMIT 2015
JACKSONVILLE, FLORIDA
DECEMBER 1 - 2, 2015
SAVE THE DATE
FDOT
fice
FLORIDA AUTOMATED VEHICLES
FDOT
www.AutomatedFL.com

Connected mobility and automation in London

State and City Level Issues
Automated Vehicles Symposium 2015

Leon Daniels

Managing Director, Surface Transport, Transport for London

MAYOR OF LONDON

TRANSPORT FOR LONDON
EVERY JOURNEY MATTERS

London's economic context

Global Context

- Most estimates rank London in the top 10 of world cities by GDP
- London is Europe's first 'megacity'
- Majority of global Top 250 Companies European headquarters are located in London

UK Context

- London's GDP is greater than the other 10 largest cities in the UK

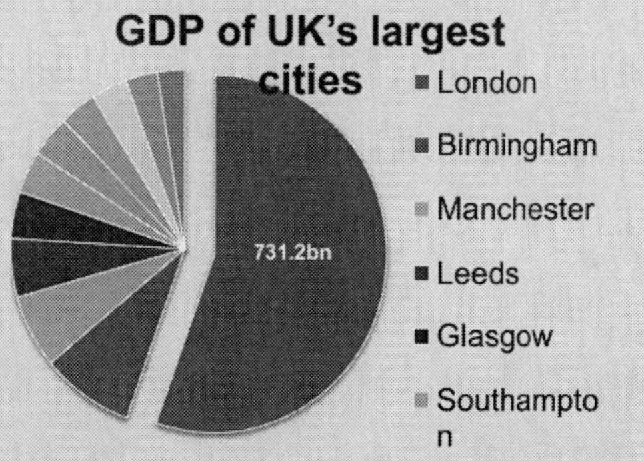

TRANSPORT FOR LONDON
EVERY JOURNEY MATTERS

London’s transport challenge – population growth

2014 2015 2030

8.4m 8.6m **10m**

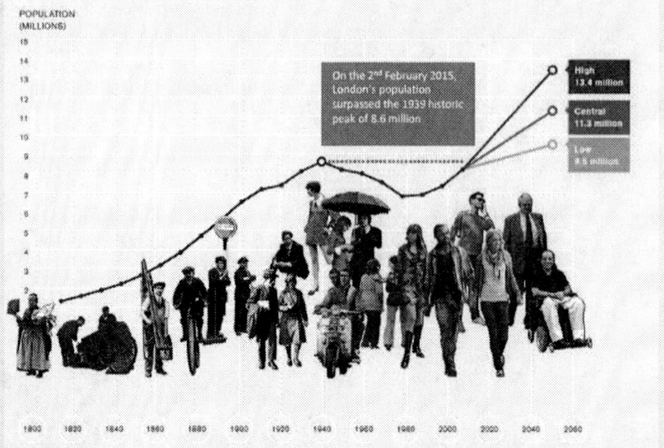

TRANSPORT FOR LONDON
EVERY JOURNEY MATTERS

What does Transport for London do?

TRANSPORT FOR LONDON
EVERY JOURNEY MATTERS

Big and Open Data

- 'Oyster' - London's Smart Ticketing system
- Contactless technology, including Apple Pay
- Integrated journey planning
- Data analytics
- 27 open data streams

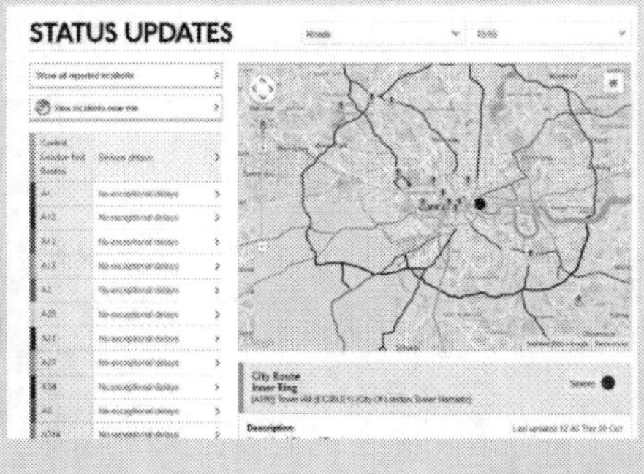

TRANSPORT FOR LONDON
EVERY JOURNEY MATTERS

Intelligent Management of the Road Network

- Smart Traffic Signals
- Real time origin destination analys
- Microsimulation tools

Image courtesy of US DOT

SURFACE INTELLIGENT TRANSPORT SYSTEMS

- Future developments
- Move to a self-regulating road network
- Enhanced data collection, processing and

TRANSPORT FOR LONDON
EVERY JOURNEY MATTERS

V2I example: London 2012 – Cooperative ITS on the network

The Challenge:
Effective road network management

- Prioritised Olympic fleet traffic

The Result: Journey times 30% faster

TRANSPORT FOR LONDON
EVERY JOURNEY MATTERS

Benefits of Connected Mobility in London

- 'Big Data'
  - Real Time Network Information
- More efficient freight & logistics
- Managing congestion
- Reduced journey times
- Mobility for an ageing population
- Non-emergencyy patient transport

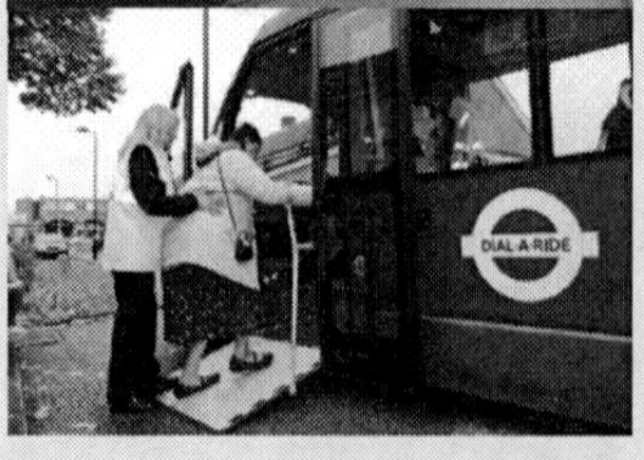

TRANSPORT FOR LONDON
EVERY JOURNEY MATTERS

Challenges for Automation in the Urban Environment

- Automation is harder
- Wide variety of vehicles & variables
- Market disruption
- Moral discourse

TRANSPORT FOR LONDON
EVERY JOURNEY MATTERS

But things are changing...

TRANSPORT FOR LONDON
EVERY JOURNEY MATTERS

Changing attitudes towards driving in London

- People are driving less
- Shared economy
- OEM mobility programmes

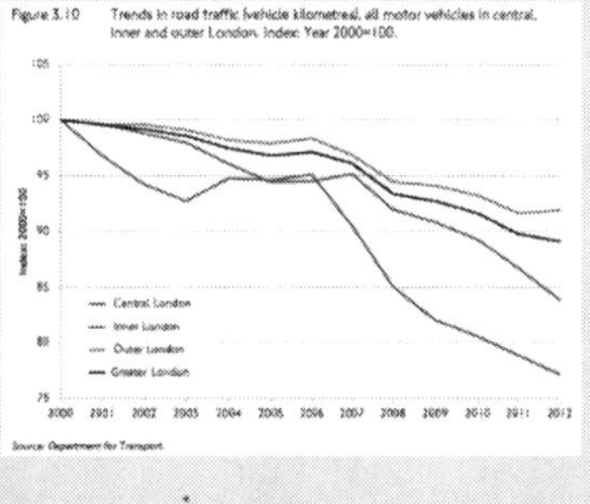

Image courtesy of Ford Motor Company

TRANSPORT FOR LONDON
EVERY JOURNEY MATTERS

A blurring of boundaries

- Taxi, Private Hire, car clubs and
- Enabled by technology and smart phones
- New entrants disrupting market
- Very cost competitive

TRANSPORT FOR LONDON
EVERY JOURNEY MATTERS

So what are we doing with automation?

- We already have automation in our transport system: Docklands Light Railway and Underground
- We have spent the last year understanding, positioning and influencing policy and strategy
- Ambitious programme for connected mobility in London: SURFACE ITS
- We have trialled V2I with OEMs and ISA on our bus fleet

TRANSPORT FOR LONDON
EVERY JOURNEY MATTERS

Testing and evaluation

- Rapid advances in technology and pace of change is accelerating
- Integrating automation into our network and services
- Risk of expensive assets rendered obsolete by new technology
- Debate over role of public authority stimulating marketplace and private sector investment and reward
- Unintended consequences

TRANSPORT FOR LONDON
EVERY JOURNEY MATTERS

Key Messages and Next Steps

- London is perfect to try innovative technologies

- Integrated Transport System

- Benefits and unintended consequences
-

- **Come and work with us in London!**

TRANSPORT
FOR LONDON
EVERY JOURNEY MATTERS

AUTHOR INDEX